building a Successful School

building a Successful School

mike WALSH

To Bernadette, Rachael and Martin, whose support has made this book possible.

I wrote this book for all those who strive to give children the very best in their education.

First published 1999

Kogan Page Limited
120 Pentonville Road
London N1 9JN

British Library Cataloguing in Publication Data

A CIP record for this book is available from the British Library.

ISBN 0 7494 3029 X

Typeset by Saxon Graphics Ltd, Derby
Printed and bound in Great Britain by Biddles Ltd, Guildford and King's Lynn

CONTENTS

ACKNOWLEDGEMENTS

I wish to thank a number of people for their help and influence in writing this book, especially Martin Cross who gave me the initial idea and encouragement to undertake the project. Thanks to Linda White for her comments and advice as the script developed. Thanks also to Jessica Saraga for her helpful comments and suggestions on the completed text. Special thanks to all the governors, headteachers and deputy headteachers I have worked with and for over the years and especially to Jim Atkinson, Sam Brown, David Ford, Trefor Jones, Brian Giffen and Chris Hopkins. Particular thanks to Patrick Hennessey whose single-minded and unselfish devotion to improving the quality of education for the underprivileged pupils at his school should be an example to all. Ian and Enid Longrigg deserve a special thank-you for the spectacular peace of Leonards Cove where I found the perfect environment for writing. Also, thanks to my editors and publishers, especially to Lee Hodder for her initial encouragement and enthusiasm and to Jonathan Simpson for his valuable comments, suggestions and advice in completing the manuscript.

Special thanks to those who contributed towards specific examples used throughout the text who, either for reasons of modesty or a desire for anonymity, did not wish to be named!

INTRODUCTION

Schools matter. As children, we attend them trustingly because our parents send us there. We know, as children, that our parents only want the best for us. We go to school believing that we are being taught to the same standard as other children in other schools. As we grow a little older, we become aware of the differences between children in our class. Some can read faster than others; some can read more difficult books; some are really good at mathematics, tackling problems and calculations that others are unable to even guess at. We notice others too: those who receive special help for their reading; those who cannot write when everyone else can. We notice those who misbehave – some children have the gift of being able to annoy the most patient of teachers.

Quite early on in primary schools, children take their place in a hierarchy of attainment, behaviour, attitude and popularity. As children progress through 11 years of compulsory schooling, there is little significant movement within this hierarchy. A well-respected reception class teacher of many years' experience has said to me on more than one occasion that she knows within months of meeting a class for the first time which children are likely to be the troublemakers and potential criminals in their teens.

When we as children do well at school, we think we are brilliant. Our school – the school we know because we spend so much of our life there – is always the best school. Doing well at that school puts us at the top of the tree, and makes us as good as anyone else anywhere – doesn't it?

Schools matter. As parents, we are becoming gradually more aware of the 'good' schools in our area and those that people think should be avoided. In some areas, as we all know, house prices are affected by the popularity of certain schools. We make judgements about schools on a range of insecure and unofficial criteria. Of course, we have league tables to help us. Do we all understand, though, that three additional special-needs children in one year can send a school sliding down its local league table? With the same teachers, the same quality of teaching and the same work set, but without those three children, the results would have been the same as in other years.

Many parents make their decisions about where to send their children by talking to others. Information gained is often based on how happy a child is and how well other parents think their children are doing. I once addressed a hall full of very happy parents. Many were immigrants and their children were the first generation born in this country. The parents were delighted that their children could read and write to far higher standards than they ever could at that age. What the parents did not realize was that their children's reading and writing was of a far lower standard than it should have been and could have been. How could they have known? Most parents only have experience of two primary schools in their whole lives: the one they attended as children and the one they send their children to. The majority of parents, even in schools that are failing, consider their children's school to be doing a good job – how are they to know if it isn't? League tables are unreliable and often opaque. Headteachers will never, ever, tell you that their school is underperforming. What parents need are clear guidelines that can help them identify success or failure in their child's school.

Schools matter. As employers, we rely on them to provide us with an articulate, literate, numerate and flexible workforce. Employers are often well placed to comment on the product of individual schools but are rarely listened to. After all, the professionals in education would say that employers have a very narrow view of educational success that does not take into account children's wider development. But employers should be forgiven for thinking that schools are where children should be prepared for adult life; and forgive them for thinking that this preparation should include uncompromisingly high standards of reading, writing, speaking and numeracy for all children!

Schools matter. Most of those in the teaching profession became teachers because they wanted to make a difference. They understand the value of the work that teachers do. Why is it then that many teachers find themselves in 'underperforming' schools with children they describe as below average? The author's early experiences as a teacher alerted him to an institutional sickness that sets in when teachers begin to doubt their own effectiveness and the abilities of their pupils. Collectively, cynical, depressed, overloaded and undervalued teachers will certainly constrain the attainment of their pupils. Conversely, enthusiastic, fired-up and well-motivated teachers can apparently work miracles. Children from particular backgrounds who would probably fail in other schools suddenly begin to exceed everyone's expectations with these teachers.

The difference between these two groups of staff is not the geographical location, the state of the buildings or the children themselves, but the management of the school. Few teachers can individually motivate themselves to rise above a collective depression – especially when it is caused or led by the management. The author has seen young, good teachers lose the sparkle from their performance during several years of teaching in a badly led school. He has also seen teachers, who would surely otherwise have failed in this scenario, being brought on to success by inspired managers leading good schools that, under different management, would have been very ordinary – or perhaps even failing – schools.

Schools matter. Being a school governor is one of the most selfless acts of community service that an individual can undertake nowadays. Governors have an enormous responsibility. They control the implementation of educational change and have responsibility for budgets and jobs. There is a seemingly endless commitment to evening and 'twilight' meetings. Ultimately, it is they who are accountable for the standards achieved by the pupils in their schools.

Despite all this, how well do governors understand their schools? Many falter if pressed hard on issues of pupil progress or attainment. Like parents, most have experienced few schools in their lives. They need to trust their headteachers and teachers as the professionals who will guide them. They understand the pressures that the staff are under and are reluctant to add to that pressure through questioning these hard-working professionals. Governors in failing schools often do not know that their schools are failing. Many governors in successful schools could probably tell you what makes their school successful, in terms of good standards and good behaviour, but could not tell you why the school is what it is. Many others would reply using two simple words – 'The headteacher'.

Schools matter. This book explains in layman's terms how the warning signs indicating that a school is in decline can be recognized and how the right action at an early stage can correct the decline before the problems become insurmountable. The examples given throughout the book are rarely specific to one school or one teacher. Few schools have a monopoly on good practice and few schools ever make an original mistake. The examples do, however, provide a focus for analysis of the factors that make some schools successful and leave others wanting.

The book is essential reading for parents, governors, teachers and school managers and anyone interested in understanding how gradual changes of circumstance and combinations of events can turn seemingly successful schools into schools that are failing to educate their children properly. By recognizing these features or occurrences, action can be taken to counter harmful trends in the management, teaching and ethos of the school before they begin to compromise the school's effectiveness.

Schools matter. Children only get one chance.

Part 1

Success or failure: understanding the difference

1

REASONS FOR FAILURE

In this opening chapter we will examine the areas where opinions are often formed when deciding whether a school is successful or whether it is failing. Low standards, poor progress, poor teaching, a threatening environment and poor management would be a nightmare scenario in any school. Do these descriptions tell an accurate story of a school's true performance, and is it that easy to identify weaknesses in some of these areas?

Let us look in turn at each of these potential reasons for failure.

LOW STANDARDS

Early identification: an opportunity to label or a cue to support?

The most common factor in identifying a weak or failing school, and the one often assumed to be the most important, is low standards. On examination, however, the differences between pupils' attainment in two similar schools are not always apparent to the untrained eye.

For example, the range of the vocabulary used, the clarity of the pupils' writing, or their ability to write in different styles, could all reveal subtle differences. In Year 2, children should be able to write simple reviews of books they have read, write stories that include dialogue between characters, and write simple poetry. In Year 6 children should be able to write book summaries, construct a sequence of poems connected by a particular theme, or write extended stories. If weaknesses in different aspects of their work are not addressed, it could leave children potentially disadvantaged as they move through their school.

This disadvantage can lead to lower national assessment test scores and other test results. This in turn can lead to a lower level of self-expectation and self-confidence from the affected children. Sometimes it can even have an effect on the expectations of these children by successive teachers who, in weaker schools, may consider the children less able than they actually are. This can eventually lead to lower public examination passes than they otherwise might have gained. It may, in turn, affect their chances of achieving a place on the right course in higher education.

Mistakes made in a person's education may eventually affect that person's career or job prospects.

It is not simplistic to argue that the way children are taught at any one stage of their education can affect their future. It is fact. Often, a setback in attainment can appear to be recovered by better teaching in a future year. But how much do the newly attained standards reflect a child's true capabilities? What would have been achieved if attainment had been appropriately high through the child's earlier years?

What are the differences that can affect standards? It is not enough for parents or governors to be told that 'pupils attained the national average' or 'pupils attained above average for schools in this area' in English or mathematics. Parents and governors need to see what lies behind these statements. Let us examine the relevance to the individual child of statements such as 'average', 'below average' or 'above average', and let us also consider what such statements really tell us when they are used to describe the performance of a school.

What do we mean by 'average'?

Statements describing pupils or schools as 'above average', 'average' or 'below average' warrant close consideration. By definition, some schools are bound to attain above the average and some are bound to attain below the average in order for there to be an average at all. Parents, governors and teachers are involving themselves in meaningless self-congratulation or soul-searching if they do not understand fully what any statement, used to describe the attainment of their children in relation to the average, actually means in terms of what children at their school really know, understand and can do. Considering only the actual level of attainment as an indicator of educational success is unjustifiable because children have different starting points at different stages of their education. True measures of success measure attainment from these starting points.

Consider the case of primary schools where children arriving in the Reception year demonstrate very low levels of literacy and numeracy and very poor social and communication skills.

In one school, in the heart of a socially deprived area, the children come from backgrounds where there is little self-discipline and where learning outside school is not valued or encouraged by families. The children's vocabulary is shaped by the cartoon and teenage channels of satellite television and by the street. Outside school, the pupils have little exposure to the adult vocabulary that they will need if they are to move on in life.

The school, though, recognizes the potential of these children and can see behind the deprivation and social disadvantages. The teachers work hard to move the children on. Some of the truly most rewarding work in education is in schools like this because the children make impressive gains in literacy and numeracy as they progress from year to year.

Yet in some good primary schools serving children like those described above, there are cases of overconscientious headteachers and staff who, having never actually analysed their own performance properly, are near to the point of apologizing for letting these children down. A hard look at the evidence, showing actual attainment of individual children over a number of years, may well show that the attainment, whilst lower than average, is improving consistently and sometimes significantly. The children, despite attaining low standards against a national average, may be being taught in very successful schools.

Their teachers, and the wider public, may not be aware of this, however, because they look only at the final outcomes (the national test results in primary schools or General Certificate of Secondary Education (GCSE) and other accredited results in secondary schools). Instead, they should consider the bigger picture that shows how much pupils have advanced since starting school.

There are also many schools where children arriving in reception demonstrate high levels of literacy and numeracy.

In some schools, perceived by the local community as successful, children arrive in Reception classes having learnt basic social skills at playgroups and nursery schools, and many can write their own names. They are developing early reading skills. They possess books at home, and as they grow up their parents encourage their interest in books, read to them, and encourage them to read themselves. They watch television, but this may be rationed or controlled. They are spoken to and engaged in conversation. Homework is checked by parents, and they are helped and encouraged to do it better.

> Teachers, seeing the pupils' high levels of attainment, confidently take credit for those achievements.

Alarmingly for supportive parents, an examination of the evidence can show that in some seemingly 'good' schools the attainment levels of pupils over time are not consistently rising – even though they may be attaining 'higher than the national or local average'. Difficult as it may be for their teachers to accept, these schools can be failing their pupils because the outcomes are not telling the whole story. In these instances, the true level of the school's effectiveness is being obscured because of the high abilities of the pupils on intake and their favourable social circumstances.

Of course, there are schools with favourable intakes that do indeed achieve impressively high standards, and that can demonstrate, even given advantageous starting points, good gains in knowledge, understanding and skill. There are also schools with disadvantaged intakes whose children fail to demonstrate appropriate gains. These children, disadvantaged in society before they start, are among the worst affected when schools fail to educate them properly.

Defining the 'average' school

Let us look now behind the statements at some of the detail. Overall, results in English and mathematics, as presented to parents and sometimes to governors, can hide a multitude of inconsistencies. For example, average English attainment could be achieved on the back of good reading and good speaking skills; however, pupils' handwriting may be below average and their ability to write in different styles (narrative, conversational, poetry and reporting, for instance) may also be below average. Deficiencies in individual areas can take years to put right – and all that time the true potential of the affected children is being blocked because they cannot write as clearly as they should be able to, using the right vocabulary for their age.

Individual schools tend to develop their own areas of strength and weakness. If schools have not successfully taught the skills of writing to their pupils in the past, it will require an enormous and co-ordinated effort by their teachers to do so in the future. But first these schools have to recognize the problem. Parents trust the information they are given by teachers, but often this information is very general and not given in any context. Only by looking in detail at, and understanding, published results – and unpublished data, which is often available on request from schools – can parents and governors discover the true extent of a school's success.

It can be argued that for a school to be able to say it is achieving 'above average', or even 'average' results in a subject, then attainment should be correspondingly 'above average' or 'average' in every aspect of that subject. This gives a much clearer picture to governors and parents, showing exactly where in a subject the school's successes are and whether there are any weaknesses. As national assessment results are one factor used to measure the effectiveness of schools, an analogy with a car's MOT test would not be inappropriate: if a car passes on almost all points but fails on the brakes, then the car quite simply (and correctly) has failed its MOT test; it is meaningless to argue that the car *on average* passed its MOT, because without brakes it is unfit to drive. Similarly, if a school is not teaching its children to satisfactory levels in one aspect of English, it is wrong to argue that English teaching is average overall, however well the children are doing in the other areas. An overall average score may mean average reading, average writing but poor spelling. Try convincing the employers who are worried because poor spelling by young employees is damaging their company's image that the schools have done a satisfactory job because the children attained 'average' results overall. Like the car with faulty brakes at its MOT test, employers will be justified in saying that if new employees cannot spell proficiently, then they have been failed by their school regardless of how well they read or use language in their writing.

So, governors and staff in schools where attainment is average, or even above average, must not be complacent. How do they know that the level of attainment is doing justice to the capabilities of the children in their school? Are they sure that attainment is consistent across all aspects of the subjects? Equally, parents in an area where attainment in the local school is below average do not necessarily have to be concerned: some children may be attaining good levels considering their starting points, and others may be demonstrating average or higher-than-average levels of attainment.

Whilst standards are one indicator of a school's performance, they are not – and should not be – the sole indicator. Outcomes alone cannot be the deciding factor in determining the quality of a school. That is to deny success for thousands of children who achieve despite low starting points and disadvantageous backgrounds. It is to deny success to teachers who could teach to much higher outcomes for less effort in more favourable circumstances. It is to deny a local community the pride it should have in knowing that, despite its location, its school is one of the best.

Summary

When trying to identify the characteristics of a failing school, the one attribute that would be most people's first choice, low standards, is found to be unreliable. There have to be far more valuable and relevant ways to measure a school's true level of success than simply looking at the end results.

POOR PROGRESS

To find the real measure of a school's success, we have to go behind the standards attained and consider the route the children took to get to their particular level of attainment. The measure of pupils' progress as they move through the school is the single most important indicator of a school's effectiveness in educating its children.

As we have seen, because a school scores well in national testing and in local league tables, it does not necessarily follow that the school is successfully teaching its pupils to the best of their abilities. Similarly, before writing off the schools lower down the league tables, it is worth taking a closer look at what they are actually achieving with the pupils they have on their roll.

The search for 'added value'

Measuring children's progress between two or more given points can determine whether a school is truly adding value through its teaching. For example, it is intended that in English, mathematics and science, the core National Curriculum subjects, children will progress by one level every two years. Many schools now keep good records to track pupils' progress between national assessment tests held at the end of Key Stage 1 and those held at the end of Key Stage 2. In order to set meaningful targets for future improvement, such records are essential.

Take, for instance, a school where in broad terms 80 per cent of children attained Level 2 in English at the end of Key Stage 1, 12 per cent gained Level 3 and the remaining 8 per cent gained Level 1 (see Figure 1.1). Four years later, if the school is to be deemed to be doing a satisfactory job with its pupils then, correspondingly, 80 per cent should have achieved Level 4, 12 per cent Level 5, and 8 per cent Level 3: the class in question should have moved on two National Curriculum levels in the four years. If, however, only 70 per cent were to achieve Level 4, 6 per cent achieve Level 5 and 24

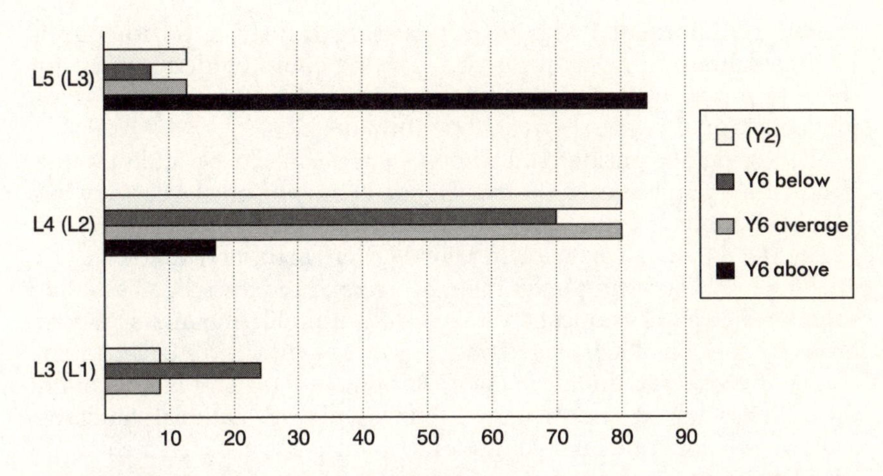

Figure 1.1 *Year 6 and Year 2 attainment for same cohort*

per cent achieve Level 3, then the class has not moved on in accordance with expectations, indicating poor progress and underachievement. Conversely, if 84 per cent were to attain Level 4 and 16 per cent attain Level 5, then the school is demonstrating good progress and high achievement.

However, life is rarely that straightforward and this kind of simple comparison is notoriously unreliable. To be absolutely certain of its performance, a school needs to track individual pupils' progress rather than global performance. The measure of one National Curriculum level over two years is only truly relevant if applied to individual children. This is because some schools face a significant turnover of pupils throughout Key Stage 2. There may be good reasons for this, including local employment patterns or parents moving children to the independent sector, which may not necessarily indicate a lack of parental confidence in the school. The school may well then take in other children who may not be attaining at the same levels as those it has lost, and a cursory look at the performance statistics between the end of Key Stage 1 and the end of Key Stage 2 may then show a decline, even though that decline may be due solely to the influx of less able children in the last two years of Key Stage 2. Remember that, in a class of 30 pupils, three pupils constitute 10 per cent of that class. If the three who leave were attaining at Level 3 at the end of Key Stage 1 and the three (or more) that replace them were to attain only at Level 1, the effect of that change in a school's position in the local league table will be significant if the rest of the pupils make only the average, expected, progress by the end of the key stage. If up-to-date and accurate records showing attainment levels of pupils were transferred with pupils moving schools, a measure of added value could be determined for these pupils as well. This would help verify the effect of such pupil movement on the school's overall performance.

All these points considered, a school can be deemed to be 'adding value' if individual pupils exceed the expectation of moving one level every two years. For example, a pupil may attain a good Level 2 in English at the end of Key Stage 1 and go on to attain a good Level 5 at the end of Key Stage 2. This pupil is progressing faster than the average and the school is adding value to that pupil's education. Conversely, a pupil who attains a good Level 2 at the end of Key Stage 1 and goes on only to attain a Level 3 at the end of Key Stage 2 is underachieving. It should not matter whether or not the child has moved schools in the interim as these are national standards; somewhere along the line – illness and absence apart – such a child has been let down by one or more schools.

National assessment results are not the only way of measuring pupils' progress. They are one very accessible way, when used properly and with other indicators, of determining whether a school is failing its pupils, particularly if the first set of data is an accurate reflection of pupils' actual abilities. National testing on entry to school at the age of five, known as 'baseline assessment', will enable teachers to measure pupils' progress from the moment they enter statutory education through to the day they leave. Many schools use a battery of tests at various stages throughout a child's schooling. These tests can measure aspects of development, such as reading or cognitive skills. Used properly, these tests should support the findings of other tests and can be valuable in confirming particular strengths or weaknesses in a child's individual performance.

Convincing the parents

Some parents appreciate knowing where their child stands, in terms of attainment, against the rest of the class. Nevertheless, the trend over the last two decades to give descriptive reports with general grades for effort and attainment has not reassured parents that schools understand the very specific needs of their individual child.

Despite the fact that many schools use a range of tests to assess attainment, few parents receive the data collected from these tests and even fewer receive it in a form that is easily understood. The schools – and they do exist – that can present parents with individual charts showing their child's reading scores, National Curriculum levels, and other indicators to demonstrate progress over time, and to show where the child is in relation to the overall range of performance in his or her class are serving parents very well in terms of sharing information. Often, even governors do not get this quality of information; but parents can and should request it from schools. Given adequate notice and assuming that the school has assessment and recording systems in place, the information should be available.

Figure 1.2 shows a child's reading age in relation to her chronological age, compared with others in her class. The chart shows that Maria M's reading age, at 23 months ahead of her chronological age, is good. However, in the context of the whole class her reading age is only slightly above average, with some children being three to four years ahead of their chronological age.

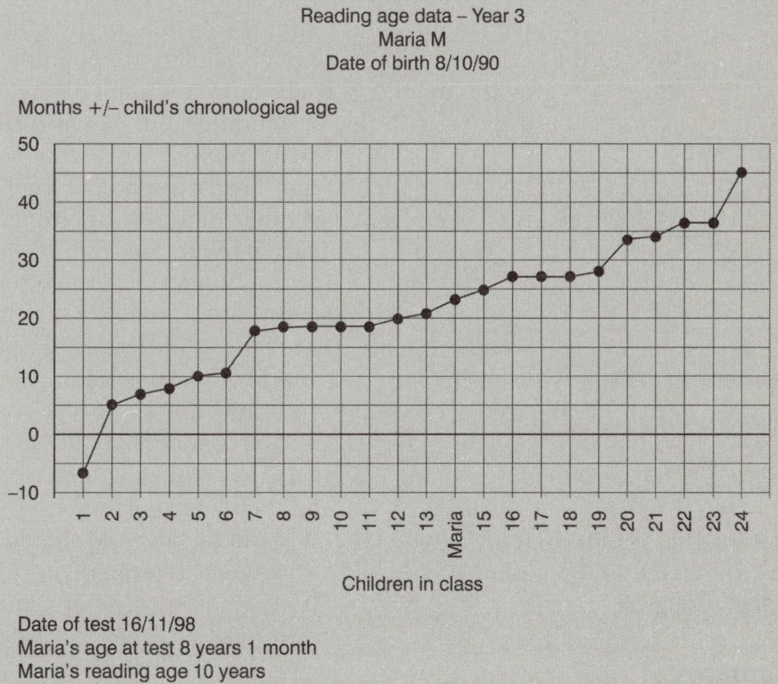

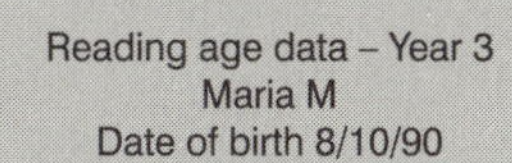

Date of test 16/11/98
Maria's age at test 8 years 1 month
Maria's reading age 10 years

Figure 1.2 *A child's performance in relation to her class*

One other thing that parents can do, and schools should do more, is to keep children's work and evaluate progress through direct comparison – for instance, a comparison of this year's handwriting against last year's handwriting, or the range and difficulty of the mathematics being studied this year as compared to the previous.

One of the most valuable sources of evidence for school inspectors during the statutory inspection of schools is an exercise called the scrutiny of pupils' work. This involves taking a sample, representing average, below-average and above-average pupils in each class, and looking at it in detail in order to check the range of work covered, the standards demonstrated, the quality of the work, and the progress made over time. In some failing schools, it is possible to see that pupils have not progressed at all; indeed, sometimes regression is clearly evident between a pupil's work in one year and that same pupil's work in the previous year. The pupils have not only failed to learn in accordance with their abilities but they have actually begun losing some of their previously acquired skills and knowledge!

Teachers should in such circumstances spot the decline in the quality of handwriting or the lack of development of vocabulary (depending on how the failure or regression manifests itself) but they don't always do so. Headteachers should spot the decline in standards, but they probably don't have systems in place for monitoring pupils' work this way. Parents might have spotted the trend had they been presented with their children's work at parents' evenings or at other times. But going back through pupils' old work is tedious for teachers who are, rightly, very concerned with the work being done *today*; but, if teachers are going to be confident about the progress that their pupils are making, the task needs undertaking. Some schools therefore undertake periodic sampling exercises, whilst others use old and new work to demonstrate progress to parents at parents' consultations or meetings.

Schools without appropriate systems cannot be confident about the real progress that their pupils are making. National Curriculum level descriptors are, despite redefinition, relatively broad, and their interpretation needs to be relayed to parents in terms of what the child can do now and should do next, rather than a description of the child's attainment in terms of any kind of stark number.

Summary

The progress made by pupils is a true record of a school's success. There are many ways to measure a child's progress. Some, such as reading tests or National Curriculum assessments, measure against clear national criteria for age. The most valuable and accurate strategies used by schools include a combination of these and a consideration of progress evidenced through analysis of work over time. Good schools will use these strategies openly and share the whole cohort's results with governors, as well as individual pupils' results with their parents.

Weaker schools will rely heavily on statutory testing and will not undertake the kind of analysis necessary to present these test scores in a meaningful way. Parents should be concerned if a teacher or school, given reasonable notice, cannot provide an accurate assessment of a child's capabilities and the progress made over time, as evidenced from a number of sources including the child's own work.

POOR TEACHING

Schools should never be deemed to be failing if the teaching within them is good. Conversely, poor teaching undoubtedly puts a school at risk of failure.

Whereas the progress that children make is the most crucial factor in determining whether they are being served well by their school, this progress is almost totally dependent upon the quality of the teaching that the school provides. Time after time, published inspection reports show that good progress occurs when there is good teaching, average progress occurs when there is adequate teaching, and slow progress occurs when there is weak teaching.

So what is so special about teaching?

Firstly teaching is a much more complex activity than the average businessman, parent or politician would appreciate. People see children's long holidays, and short working days and imagine that, on the face of it, working with children as a teacher looks like a pretty easy way to earn a living. Little could be further from the truth, as those spouses married to teachers would be the first to agree!

The days are not short. The children may only be in school for six hours or so, but the teachers are rarely in the building for less than eight or nine. We are all familiar with the teaching unions complaining about the hours worked at evenings and at weekends as well. It is a shame that the general public does not consider their comments representative of the truth, because the way the teaching unions put them over does not chime with society's expectation of teachers being reasonable, professional and caring individuals. The public image presented by teachers' unions at conferences does little to inspire credibility. However, it is true that the majority of teachers – and certainly the successful ones – put in ridiculously long hours at home each evening, every weekend and for good periods of their holidays, in order to research, prepare and mark work and to keep records up to date. And all that before they start writing schemes of work or policy statements.

Consider the children. There are, in most classes, 30 or so of them with one teacher. In a primary school, the children are usually lively and not always well disciplined socially; the teacher needs to be 100 per cent on the task throughout the day to ensure that the time the children spend in school is spent working and learning. There is absolutely no chance of walking away from a difficult task for a few minutes to make a coffee and reconsider your strategy. When the children are there, the teacher has to be there, and the strategy has to work – hence the time put in the night before on planning. In a secondary school, the children are a little less forgiving. During the course of a day, a teacher can see 100–150 different children, making it difficult to establish a meaningful relationship. Under these circumstances, if the lessons are not of good quality, interesting and relevant to the pupils, there is little loyalty or tolerance and pupils' concentration and behaviour is seen to begin to slip.

The boredom factor sets in easily in primary and secondary school classes, and the teacher needs to be absolutely on top of the class at all times, with work that challenges the children properly. If the work the pupils are given is too challenging, then the pupils will not be able to do it and there is a risk of them becoming restless; if there is too little challenge and the work is too easy, they will become bored and, again, there is a risk of them becoming restless.

So what makes *good* teaching?

For teaching to be really effective, several factors need to be present. A good level of knowledge about the pupils' previous attainments, abilities and attitudes to work is essential. The teacher should also have good personal knowledge and understanding in the subject being taught. Personality and presence are essential, as is the ability to organize learning effectively and use time well. The teacher's character, in particular the way he or she talks to pupils and other adults, is also a significant factor. Here are some pointers to illustrate good teaching:

- ❑ *Good teaching starts with teachers knowing their children.* A good teacher will know what each child in that class has already achieved and what such children are capable of learning next. This requires a knowledge of attainment and progress to date. Before a good teacher can even begin to plan a good lesson, there have to be adequate records showing what has been covered so far and how well individual children have acquired the necessary skills, knowledge and understanding in the subjects to date.

- *Good teaching requires a lesson plan that ensures that there is something in that lesson for everyone*, regardless of their previous learning. There are teachers – indeed there are whole schools – credited with being 'good' by parents because they teach the more able extremely well – but they do not do a particularly good job for the less able pupils. Most parents of less able children do not complain too loudly because they are uncomfortable about criticizing a seemingly successful system, and this despite the fact that their children may be coming home miserable every night and feeling more and more inadequate in their own capabilities as the months in that class, or years in that school, go by. Similarly, there are teachers and schools that are very good at teaching the less able but do not tend to stretch the more able. Parents of the more able feel uncomfortable about complaining; they are not quite sure that their child is above average and do not want to be seen as unnecessarily pushy. In either of these contexts, some children are being failed by their teachers. The fact that the needs of the majority may be being met is simply not good enough. All children only have one opportunity to be properly educated and prepared for adult life, and this opportunity should not be wasted.
- *Good teaching requires the teacher to have high levels of subject knowledge.* If a teacher is uncertain about the mathematics he is teaching or the purpose of a science experiment she is going to undertake with her class, then that uncertainty is almost sure to be transmitted to the pupils. Teaching is about confidence, and learning is about gaining confidence. A teacher who lacks confidence in the classroom is not going to be successful in helping children learn. Pupils will not always understand the point being made, and so the teacher needs to have sufficient confidence in the subject to be able to manipulate the subject content to suit the pupils' current levels of understanding.
- *Good teaching requires presence and personality.* It is amazing how many seemingly shy and rather withdrawn adults there are who, once in the classroom, can manipulate a class full of children through humour, good storytelling, a look, and the impression that they are totally in control. These attributes come with practice to most teachers and are best expressed when combined with good subject knowledge and a clear understanding of what exactly the pupils are being required to learn.

- *Good teaching requires good classroom organization and time management.* Lessons need a beginning, a middle and an end – preferably not interrupted by the morning or lunchtime break. Children need access to the resources that will help them in their learning; they need to see the board; they need to see the teacher; and the teacher needs to be able to see them.
- *Good teaching requires the teacher to be a good role model.* This is reflected in the way teachers talk to children and deal with problems that arise during a lesson. It is reflected in the quality of teachers' worksheets and board work. It is reflected in the teachers' attitude towards their own work and their attitude towards other adults.

Some of these qualities can be learnt whilst others have to develop over time. Some can be developed through training whilst others are to do with a teacher's own perspective on life. In schools where teaching is deemed to be poor, many or all of these attributes are lacking and their absence directly affects the chances of pupils making good progress. So how can parents or governors tell if these qualities exist at the school they are concerned with? Here are some negative pointers:

- *Poor planning usually results in unfocused teaching.* If lessons are not being planned, then learning will not be properly supported. Parents of primary-age children can ask themselves whether they receive, or have access to, a summary of the work being done in their child's class during a particular term; the same can apply to secondary-school subjects. Whereas the summary alone does not guarantee good planning, its presence does indicate a level of organization that could lead to good planning. Good schools display the medium-term (half-termly or monthly) planning on the wall of a classroom for parents to see; others make this planning available on request. Good schools should be able to tell parents that the headteacher sees short-term (daily or weekly) planning every week and that this planning is carefully checked to ensure that it follows on appropriately from the previous week. It is surprising how many schools do not have such systems in place, especially when the systems would give parents extra confidence in how their children are being taught.
- *Poorly motivated teachers cannot motivate children.* If children are coming home with tales of noisy lessons and regular poor behaviour and, as a parent, you cannot find evidence of quality work in their school books, then it is time to make enquiries. In the short term a child may think it amusing that school can be such good fun, but in the longer term the child's education may suffer irreparable damage.

- *Poor lesson or classroom organization will undermine otherwise well prepared lessons*. This situation is particularly the case if these factors are not properly taken into account at the lesson planning stage. Classes are likely to be noisier and more difficult to control and children's concentration and learning is likely to be interrupted, often without the teacher realizing or appreciating that it is happening. Learning and progress will be significantly hampered.
- *A poor role model will have a negative effect*. Teachers' personalities and the way they conduct themselves in the classroom impact directly on the values and qualities that contribute to a child's whole-life experience and the way that a child will react to other people and other challenges in the future. In the days before the National Curriculum, it was often said that we remembered little of what we were taught as children at school but we remembered the good and the bad teachers. These are the role models who shape our lives, either through emulation or avoidance of the qualities they possessed. There is still a good deal of truth in this statement.

An inspector was driving through one of the less appealing areas of a northern English city some 10 years after he had last taught there. The road layouts had changed and he could not find the school he was looking for. Whilst struggling to read an old A–Z and relate it to new road layouts, he became aware of two large and rough-looking youths peering through the window of his unlocked car. Apprehension turned to relief and happiness when they asked if he remembered them. He didn't, having only vague memories at best, but they remembered him and his firm but very fair control of their very tough class. They remembered the fun they had had in the teacher's music lessons, but they also remembered how hard they had had to work to enjoy that fun. Although neither of them went on to study music in more depth, they were very genuine in their appreciation of the lessons taught by this teacher. And they had learnt that hard work and enjoyment are not necessarily exclusive.

Governors, parents and the quality of teaching

Governors can gain some idea of the quality of teaching by visiting a school and walking around the various classrooms. Which are the organized ones where there is an air of purposeful working? Which are the rooms where the teacher is confident, in control and has the attention of all the children when required? Which are the rooms where there is high-quality work displayed on the walls and observed on pupils' desks or tables? Governors should not discuss their observations with the teachers directly but should share them with the headteacher and seek professional clarification and interpretation of what they have seen. Governors can then develop their understanding of the headteacher's management role and provide appropriate support to the headteacher in fulfilling this role.

Parents can derive clues about the quality of a teacher's work by looking carefully at their children's homework and exercise books. Has the homework been clearly explained to their child? Was the previous homework properly marked? Did you agree with the marking? Were there any comments written by the teacher? Were those comments meaningful to your child? Is the work well presented? Does the teacher expect it to be well presented? Parents have the right to ask their children's schools about the way homework is set and about the standards of presentation that the school expects. They should be prepared to accept a good rationale for the former and the schools should be prepared to accept fair criticism of the latter.

Only a supply teacher?

It is worth remembering that all the expectations of good teaching should be applied to supply teachers and student teachers as well as the regularly employed teachers in the school. It is wrong for a headteacher to blame a class's slowing progress or deteriorating discipline on a supply teacher. The headteacher employed that supply worker and the headteacher is ultimately responsible for his or her performance. Good schools will brief supply teachers properly on the work to be done and on the school's expected standards, and will then monitor the supply teacher's performance. Good supply teachers will not mind this: they often survive on their reputations and will appreciate someone noticing their successes in adapting to a particular school's ways.

The work of student teachers should be carefully monitored at the preparatory stage as well as at the point of delivery. The school has the right – and good schools exercise that right – to refuse to let clearly failing student teachers back into the classroom. Governors and parents should all be satisfied that the school's expectations of supply and student teachers are appropriately high and that these professional visitors to the school are properly supported.

Summary

Governors should know that their teachers are good teachers, and parents should feel confident that their child's class teacher is looking after precise and individual needs. When schools fail, it is because teachers are not addressing the children's needs appropriately; because they are not sufficiently knowledgeable about what it is the children know and should learn next; or because they do not offer appropriate models of commitment, fairness and integrity to children in their formative years.

A THREATENING ENVIRONMENT

Sometimes schools fail because they are not places where anyone who knows and cares about what goes on inside them would wish to leave their children. These schools usually demonstrate low standards and poor progress. The deteriorating environment stems in no small part from a lack of rigour, discipline and expectation.

A threatening environment is not the sole reserve of the secondary-schools sector. Although the extreme cases that make the national headlines are usually secondary schools, there have been one or two notably unpleasant descriptions of unsafe environments for children in primary schools. The ones making the headlines are the easy ones to spot. They are the schools where an extreme incident or series of incidents, ranging from violence against pupils or staff or well-publicized breakdowns in management and control, make easy and sensational headlines. However, there are other schools where the environment can be just as uncomfortable but there has been no well-publicized catalyst to propel that school into the national domain.

A threatening environment usually means that bullying, and often racism, are regular occurrences, resulting in children being frightened, corrupted into taking part, or both.

Bullying by children

The first consideration in determining whether a school is a secure environment is the happiness of the children. Bullying occurs in every school, despite what the headteacher may say in the school newsletters or at the open evenings. The biggest problem with bullying is actually defining what it is. There is now a growing acceptance that bullying taking the form of taunting and teasing is just as damaging as physical bullying, if not more so. It is certainly more difficult to spot. Then there are degrees of teasing or physical action, and the identification of the point where these actions become bullying. Even gentle teasing may be upsetting for some pupils in much the same way as a gentle push may upset others. Other pupils may not consider themselves bullied unless taunted mercilessly or punched.

The problems in defining bullying

In one primary school a group of boys and girls was asked if there was bullying in their school. The girls replied 'yes', because there was a lot of cruel name-calling. The boys replied 'no', because they thought that bullying only occurred when someone was hit, hurt and didn't deserve it! This school had already undertaken what it thought was a specific and successful teaching programme to tackle bullying, but still these old perceptions remained unchanged once the pupils left the classroom.

Many schools now have anti-bullying campaigns supported by a plethora of initiatives such as peer mediation, pupils' courts and counselling. Parents should ask their children's school what initiatives it is taking to counter bullying. If parents are told in response that there is no bullying and therefore no initiatives are needed, then they need to dig deeper. An equally worrying response is if they are told that all children tease and get teased and that playgrounds can be a little rough. Whilst there is some truth in this statement, the school needs to set minimum standards of behaviour for children, and the school needs to have ways of promoting and enforcing these minimum standards. It is also essential that parents actively support their children's schools in any initiatives to prevent bullying.

Whilst children will only sometimes admit to being bullied themselves, they will almost always say when asked when they know someone else who is being bullied. This is especially the case when they consider that a great injustice is being done to the victim. They will also say when they feel that bullying is being dealt with effectively.

Bullying by adults

Pupils in some failing schools are not only at risk from other pupils but their teachers too can present a physical or emotional threat to their well-being.

In the same way that it is possible for adults to bully other adults at work, there are sadly also occurrences where *teachers* bully children.

Headteachers need to know that the respect paid to each individual within their school is of the highest order. Children learn more than the National Curriculum from their teachers, and in schools where there is serious bullying within the pupil community it is worth looking to see how the adults in the school treat the children. What values are they teaching them? Cynicism, sarcasm and ridicule – too often featured in the educational experiences of children taught earlier in the century – have no place in a modern teacher's repertoire of class-control strategies. Children who are the subject of an oppressive and sometimes mentally cruel regime, in a school or in an individual class, are more likely to behave irrationally to others than those who are taught in a regime where there are clearly shared values and expectations and consistent and fair responses to misdemeanours. It is possible to find some individual teachers behaving this way in otherwise good schools. Children should not have to be taught in these circumstances in any school.

Racism

Racism in schools, between pupils, is another area where it is quite difficult to dig out the truth. Few Ofsted (Office for Standards in Education) inspection reports will say that there are occasional racist incidents in a school, because the trouble that this could cause for the school in the longer term may far outweigh the benefit of publication of such a statement in the first place. This approach, however, does nothing to take away the hurt and upset caused to victims of racism who may feel that the school is trying to cover the issue up and ignore their problems.

Few headteachers will say, outside professional circles, that there is racism in their school. Those who do are to be applauded. Talk to children, though, and the reports of racist-related bullying can be frequent and, in some cases, worrying. It is not simply the white–black scenario that is the most popular arena for racism; Asian children, for example, have described racism between themselves and others from different parts of the Indian subcontinent. Any minority group is a target, and refugees are often targets for bullying by white, black or Asian majorities.

Governors need to be sure that their schools have up-to-date and well-thought-out behaviour-related policies on racism. More than this, they need to know that these policies are effectively practised and are successful in addressing and eradicating antisocial behaviour within the school boundaries. Parents too need to know that the school has procedures for promoting positive behaviour and for consistently tackling and countering racism.

Teachers can be 'at risk' too

It is not only children who are at risk in schools that are failing to provide a secure and stable environment. In some schools, teachers and other adults are very much at risk. Where there has been a breakdown of discipline in a school and where children believe that they are the ones who are actually in charge, it only takes a small incident, or a child with something to prove, to involve a teacher or other adult at the school in an act of violence. Violence towards teachers by pupils is well documented and occurs more often than parents probably realize; and it does not occur solely in secondary schools. Many teachers have been injured by unruly primary-school children of varying ages.

Governors and parents need to know that their teachers are safe and that violent incidents are not common in their school. Parents cannot be confident of the safety of their children if their children's guardians are fearful for their own safety.

Knowing that a school is 'safe'

Whilst a whole school demonstrating the above shortcomings may be relatively easily identified by a curious and determined parent, an individual teacher demonstrating these kinds of weaknesses within a school, or within an individual class in a primary school, is not so easily identified. However, these teachers and classes can seriously interrupt a child's educational and emotional development. There are schools where parents know that children in a particular class get a rougher deal than in others. There are schools where parents know that bullying is more common in some classes than in others. Some headteachers suspect this is the case but often the teachers do just enough to stay on the right side of a disciplinary enquiry. Other headteachers choose to ignore the situation, quietly looking forward to the day when the teacher retires or gets a job elsewhere.

Often headteachers will say that they cannot take action against a teacher unless parents write to them with a specific complaint. Often parents will not wish to do this because they fear things will be made worse for their child if they do. But what kind of a response is this? Things are already bad for their children, who are the ones who will have to tolerate the excesses of that teacher for a whole year. Parents should write those letters – politely and firmly and with very specific details. Parents should say when they expect a response and should follow the letter up if the response is not received. They should reserve the right to complain to the highest authorities if their child suffers as a result of the complaint.

In reality, teachers under investigation will usually go out of their way not to make things any worse for the child of a complaining parent, and so the child may not even know that a complaint has been made. Every school should have a published procedure that explains to parents how they can progress complaints. Schools should be grateful that parents are supporting them in dealing with weak teachers and must be prepared to follow up every complaint thoroughly.

Summary

The negative effect of a disturbed environment on the progress that children should be making is usually clear. When children are not happy at school, they will not enjoy their work. If they do not enjoy their work, they will not apply the appropriate effort required to progress. If the energies of the teaching staff are being taken up by resolving potentially violent disputes and situations, then they are not appropriately engaged in properly teaching their pupils.

Conversely, if children feel happy and secure, experience good teaching by teachers who themselves feel secure and understand that teaching is about developing systems for living together as well as imparting knowledge, then these children will progress appropriately.

POOR MANAGEMENT

One common factor shared by almost all failing schools is poor management. Schools are complex organizations reliant on the successful interaction of hundreds of people every day to make them work. What then, in these complex organizations, are the elements or combinations of elements that require good management?

Areas for management

There are many aspects of a school that require professional management. These include the children, the curriculum, the teachers, the non-teaching staff, the learning resources, the buildings and the grounds. Poor management of any of these can constrain, directly or indirectly, the progress made and standards achieved by pupils. Let us look at each of these in turn.

The children

Firstly, there are no 'raw materials' or 'ingredients' to process in a school as there are in a factory, and so no two schools are working from identical starting points. Children are not 'raw materials' in that sense. Each child is an individual with a different set of feelings and needs. Their educational, social and developmental needs all need managing. This includes the classes that they are placed in and the groups and sets that they are assigned to for different subjects.

Some children will have special learning needs and may require support. Others may have skills or interests that may be nurtured through (for example) extracurricular activities in sport or music or chess. Some pupils may have particular dietary and health requirements, the latter implying training needs for certain staff who come into regular contact with them.

The curriculum

There is a common curriculum to be imparted and experienced at any stage of a child's school life, and within that curriculum there are specific skills, bodies of knowledge, and concepts to be acquired, learnt or developed. The point of interaction between individual children and the curriculum is in a constant state of change; it is often different for different children, and it needs careful management to ensure that it is properly balanced and remains relevant for each child.

The teachers

Teachers have a range of responsibilities, always beginning with the children they teach but often including year, key-stage or whole-school responsibilities for either a subject or an area (such as assessment, or health and safety). Each teacher has a different experience profile and a different set of skills and professional interests. A headteacher who uses the experience, skills and interests of these teachers correctly can simplify his own job whilst improving the effectiveness of the school and gaining a happy and motivated staff. The headteacher who ignores them loses the advantages of sensible delegation: teachers may feel undervalued, with a subsequent effect on performance and may eventually leave, thus giving a wiser and more perceptive headteacher the benefit of their skills and potential.

In addition, there are professional visitors to the school who may teach instrumental music or support children with special educational needs whose working conditions and potential for overall contribution to the school must not be overlooked.

The non-teaching staff

The non-teaching staff comprise the administrative officers, the school keepers and cleaners, welfare assistants, and non-professional support staff. All these individuals play a vital role in the smooth running of a school, and their professional needs warrant careful management. Parents and visitors will form a poor impression of a school if the receptionist is brusque, the telephone rings unanswered, or the school is dirty and untidy.

Showing parents that a school cares

A particularly well-managed primary school has a 'no appointment necessary' system for parents wishing to see a teacher. The school provides a waiting area, together with a kettle, tea, coffee and biscuits, and pupils' work is on display there. The understanding is that if parents are prepared to wait, then they will be seen as soon as it is possible.

The learning resources

A school's learning resources no longer comprise simply pens, paper and textbooks. Many primary schools have an impressive array of electrical equipment or computer-based equipment, and the new national grid for learning will, in time, revolutionize the internal appearance of many classrooms. School libraries often require the services of volunteer helpers, in addition to professional staff, in order to keep them organized and attractive. Schools undertake a number of visits, trips and occasional residential trips each year. In addition, there is usually a programme of professional visitors to the school who, when used properly, enhance the learning of the children.

The buildings and grounds

Headteachers need to consider everything in respect of the buildings and grounds, from where the children are taught through to where and what they can play at breaks and lunchtimes.

Health and safety issues have to be considered. Headteachers can and do spend time employing joiners when a wooden stairway to an outside temporary classroom begins to show signs of rot. In one school I encountered the headteacher was the only member of staff who knew how to order heating fuel; unfortunately, one winter's day the headteacher was suddenly taken ill and this coincided with the school running out of heating fuel. The absence of a simple system caused the school to close for two days, interrupting the education of a whole school and disrupting the lives of hundreds of local families who had to make special arrangements for childcare at very short notice.

Then there is the issue of security of both grounds and classrooms, both pupils and staff, especially if the latter are working on their own at the end of the school day.

Keeping the ball moving

One of the main perceived injustices about life in education is that all of the above areas need managing in all schools whatever their size but that there are more people to delegate the jobs to in secondary schools. However, more managers do not necessarily create greater efficiency, and there are many very well-managed small primary schools.

A well-managed school will be well organized. Staff will know their responsibilities and everyone in the school will be familiar with the school's routines and expectations. A well-managed school will be able to meet parents confidently and answer any questions they may have about their child's education. A well-managed school will make its parents aware of much information before it is asked for. Staff will be happy and well motivated and will always present a professional front when talking to parents. Pupils will be orderly and will know the parameters of acceptable behaviour and the standards of work that they are expected to produce. Visitors will feel welcome and will not feel that they are interrupting a busy or overworked staff.

Well-managed schools have staff who share responsibility as well as a common set of goals. They are schools where there is clear and positive leadership from a headteacher who, whilst being prepared to listen, is not afraid to make positive decisions after the necessary consultation. These schools have procedures for dealing with day-to-day occurrences. They have systems for working together and for school development. They are prepared to review their work, analyse their success rate and change their practice to improve it.

Schools where management is weak rarely have such procedures or systems. Of course, the procedures for dealing with latecomers, for example, may be written down in the staff handbook, but if the staff handbook has not been read, and has not been shared with new staff, nobody will be aware of it and use it. Parents in one school were heard to sing approval of that school because nobody cared if you were late in the morning, and so it made getting children ready for school so much less stressful!

Governors and the well-managed school

Governors have a responsibility to know that their school is well managed, and part of that responsibility is knowing that basic procedures are in place. They need to know that these procedures are understood and used consistently by all staff, whether they cover lateness, deal with incorrect spelling in pupils' work, or guide staff in responding to bullying.

It is the governors in a school who are ultimately responsible for weak management. They employ the headteacher and have the power to change the headteacher if they are not satisfied with performance. Governors in good schools have their own systems for knowing whether management is effective. These may include regular and focused meetings with management and the questioning of headteachers when they present reports at governors' meetings. In good schools, management of resources is monitored through careful budget-setting and control systems. In weak schools, some or all of these systems do not exist, for problems that cannot be identified cannot be resolved.

A school with poorly-managed teachers is easy to spot. There will be a big difference between the practice of individual teachers, their class control, the appearance of their classrooms, and the way that they relate to parents. In a well-managed school, differences in teaching styles and in the quality of the teachers' class control or the general appearance of their classroom will be much smaller and the overall consistency of teaching will be higher. If parents share major concerns about the differences in practices between several teachers in a school, such as some being stricter or setting more or less homework than others, then these are primarily signs that there are weaknesses in the management of the school and not necessarily signs of weak teachers. The teachers themselves may not be at fault but, lacking clear direction and leadership, they will be teaching in a vacuum with little awareness as to the practices and techniques employed by their colleagues.

Poorly managed pupils are even easier to spot. The absence of a framework and of consistently applied behavioural standards will result in excess noise, undisciplined behaviour in and around the school, boisterous and antisocial behaviour in the playground, and in pupils having a poor attitude to their work.

Parents have a right to expect that the people managing their children's education will be efficient and well organized. If they feel, through either a poor experience when contacting the school, through lack of information from the school, or through a perceived injustice in the way their, or another, child has been dealt with, they should make their feelings known to the school. If they are not comfortable talking directly with the headteacher about their feelings, they should approach one of the parent-governors. A letter is always useful because it makes a written reply almost mandatory. Once again, the school's complaints procedures should help parents who feel that there are significant and damaging deficiencies in the way that their child's school is being managed.

Summary

Schools are complex organizations that require good management. Weak management can threaten progress and attainment because teachers' time may be spent dealing with deficiencies in the organization of a school rather than concentrating on the needs of the pupils.

Governors particularly need to be able to recognize whether or not their school is well managed. The way parents are treated when visiting a school can be a valuable indicator of the quality of the school management.

CHAPTER SUMMARY

The five preceding sections in this chapter explain in detail the criteria used to identify potential weaknesses in schools. The main points deduced are as follows:

- Weak schools are schools where pupils are not making adequate progress rather than schools where pupils are not achieving high standards. This definition ensures a truly 'value added' criterion for recognizing success and identifying failure.
- High attainment is only really possible when pupils make good progress. Schools with a literate and 'able' intake can be at risk of failing if teachers do not recognize their pupils' abilities and move them on appropriately.
- Schools with a 'below average' intake can have teachers who sometimes assume that children will never be able to achieve.
- Low expectations of pupils, combined with poor management, are ingredients that consign generations of children to a life of low achievement and low self-worth.

- ❑ Poor discipline can result in pupils feeling threatened and insecure in their learning environment that, in turn, can depress attainment. Conversely, well-disciplined schools that challenge their pupils to succeed in seemingly difficult circumstances motivate their pupils to value their abilities, their environment and themselves.
- ❑ The quality of a school's management is a key factor in determining its success.
- ❑ Poor managers can damage children's chances of success at school by exposing them to widely differing sets of expectations and values as they progress through the school.
- ❑ Governors are ultimately responsible for ensuring that their schools are effectively managed.

2

EXPLODING THE MYTHS

In this chapter we will examine some of the myths about failing schools. The worst schools are not necessarily those in the most deprived areas, where it can be difficult to recruit staff. The worst teachers and the most badly behaved pupils are not necessarily found in what people perceive to be the worst schools. Nor, for that matter, are the best schools to teach in those in the more affluent areas. It is not necessarily more difficult to manage schools in deprived areas. All schools require good-quality management in order to succeed.

MYTH 1: 'IN INNER-CITY SCHOOLS WE CAN'T GET THE STAFF'

In 1997 the Chief Inspector of Schools, Chris Woodhead, controversially identified the number of incompetent teachers in schools as being as high as 15,000. Whether this figure is accurate or not, it does serve to underline that there are undoubtedly many weak teachers in our schools. But it should be remembered that weak teachers are not exclusively found in the tough inner-city areas; indeed, as a corollary, some of the best teachers can and should be found in areas where the job is at its most challenging.

A tale of two teachers

It is 8.30 am in the car park of a newly opened boys school in a notoriously tough inner-city area. A small sports car arrives and is reversed nervously into one of the many empty parking bays. The engine stays running. A mathematics teacher sits gripping the steering wheel, motionless and deep in thought. Other cars arrive and men and women carrying bags, books and boxes hurry to the staff entrance or directly to their classrooms. The engine of the sports car is still running and the driver is motionless. The time reaches 8.40 am and the car park is almost full. The pavement on the road outside the staff entrance is also filling up with children making their way around the building. The mathematics teacher in the sports car slips the vehicle into gear and drives away, fulfilling a ritual that has taken place daily for almost two weeks since the start of term.

One of the cars already parked when the mathematics teacher arrived belonged to a colleague who had taught with him previously at a boys' grammar school. This colleague was much younger, four years into the job instead of 24. He looked forward to teaching the children in the school even though, on the face of it, they did not have the intellectual capabilities of the grammar school children they had both previously taught. They certainly did not have the supportive backgrounds and the privileges that these backgrounds provided: these children lacked a conventional desire to learn, typified through listening in silence, concentrating hard on set tasks, and raising their hands to ask questions. However, in its place they had an enthusiasm and willingness that, if harnessed and nurtured could be used to further their knowledge and mathematical skills. Success with these children was noticeable and brought its own special sense of achievement for the teacher. This teacher felt that he was making a real difference.

Eventually the first teacher in our example was declared unfit for work and retired on a breakdown pension. By all accounts he had been a successful teacher in his previous position at the grammar school. Never a career man, he had not sought promotion. He loved his mathematics and enjoyed nothing better than sharing his favourite subject with willing and enthusiastic minds. Others could worry about the organization of the department or the taking on of additional responsibilities in order to further their career prospects. He was happy that he had a good reputation for enabling his pupils to achieve high levels of success in public examination. The grammar school he taught at had been closed to make way for this new comprehensive school serving a much wider, and much more socially diverse, area. That was the theory anyway. In practice, the school filled up almost exclusively with the inner-city children from the immediate area, and the parents from the more middle-class area where the grammar school had been sited made alternative arrangements for their children's education.

The children's enthusiasm, recognized by his younger colleague, was interpreted by the older teacher as unruliness and a lack of self-discipline. Indeed, when that essential ingredient of respect between pupils and teacher was missing, their energies were easily be transformed into objectionable and ill-disciplined behaviour. These children were prepared to accept education if they could see its relevance and if the teacher was prepared to make it accessible to them. They could not accept being required to study a body of knowledge for its own sake. No matter how many times the natural beauty of mathematics was pointed out to them, unless they could see that it had a purpose – a relevance to them – then it was unimportant.

It couldn't happen today, could it?

Time has moved on significantly since these events. We now have a National Curriculum, which is supposed to ensure that each subject is relevant for every child. School management and monitoring systems have moved on too: no longer should it be possible for a teacher to have a nervous breakdown without receiving significant support and assistance from colleagues and other agencies. And there is in-service training available for every teacher so that they are aware of the best techniques and strategies for teaching their subject.

However, these developments have not completely solved the problems. There are still many, many stressed individuals for whom facing their classes each day is a major challenge – individuals who are so finely balanced between being in control and losing it that the slightest incident can become a major crisis for them; individuals who face, daily, the taunts of their pupils and who have no response, or who have lost the will to respond, and who envy their colleagues' superior class control and ability to command respect. Whilst one must have sympathy with these misplaced characters, there is little sympathy to be had for those individuals who are in denial that they have a problem. These individuals have developed a mind-set that allows them to rationalize poor behaviour and low standards, together with the pupils' lack of respect for them, in such a way that they honestly believe that they have no responsibility for the cause of their problems or for the chaotic outcomes.

When teachers fail in a big way, there is usually someone there to notice (other than the pupils, who are always there and who always notice). An inner-city school will bring out the best in any teacher but will always amplify the worst as well. It is unfortunate for some teachers that the circumstances in which they work lead them to a nervous breakdown or retirement through ill health. It is equally unfortunate for some children that the same teachers can survive in much 'easier' schools simply because the challenges of the job are less severe. This doesn't necessarily make their teaching any better, though. It can mean that the circumstances in which these teachers are working will allow a lower quality of performance before visible problems appear.

Weak teaching – the link with poor management

Schools can be deemed to be failing by government inspectors because of poor teaching. National assessment or examination results may not show this in the short term but, if unchecked, poor teaching will undoubtedly affect a school's standing. However, slipping educational standards, measured through assessments and examinations, are not the first indicators of trouble. They are one of the last. A decline in the quality of teaching is one of the first indicators, and weakening teaching, if not quickly recognized and checked, usually indicates a decline in the quality of management. It is weak management that allows weak teaching to occur in a school, and it is equally possible for a previously good school to develop serious weaknesses as it is for a previously weak school to improve.

Schools are at their most vulnerable when there is a change of management after a run of several successful years. This management change may not necessarily be as a consequence of a change of head-teacher for the school. Whilst this may be the case in a small primary school, in a larger secondary school it may mean, for instance, a management restructuring that significantly changes lines of communication, decision-making processes and key individuals' responsibilities. More simply, it could be caused by the same people working in a very different way.

After such change, a headteacher needs to be certain that he or she can trust that the quality of provision in every classroom is secure. Some headteachers may begin new jobs in 'successful' or just 'secure' schools. They begin their new jobs with the attitude that if they change nothing. or change little, then the school will continue to perform as before. It may or it may not; in fact it probably will not. This is because schools are dynamic institutions and are very much dependent on the interactive chemistry amongst staff – and between management and staff particularly – to make them successful. No new headteacher can automatically inherit the relationship that a previous headteacher had with each member of staff. That relationship was defined over many years by the headteacher's response and reaction to every suggestion, every problem, every initiative and every incident. These responses took the school in a particular direction and gave it a certain style or character. A new set of responses will begin, from day one, to alter and develop that character in a different way.

Often, one hears a new headteacher bemoan the lack of 'systems' present in a school he or she has recently taken over. In particular, systems for monitoring the quality of teaching are rarely thought to have been present previously. It is true that in many schools formal systems, supported by documentation and carefully kept records, have in general only recent developed. Nevertheless, it is often a mistaken presumption to assume that the previous headteacher didn't have a clue about what was going on in a classroom. It is more likely that the head did indeed know what was going on but did not have a formal system that could be left for a successor. Classroom visits may have been informal but frequent, and the headteacher's knowledge of the capabilities of individual staff, acquired over time, would be supplemented through a confidential sharing with senior colleagues that only comes with the development of secure and trusting relationships. The headteacher may also have listened well to parents, having again developed a level of trust with them over a number of years.

New headteachers do not automatically have access to these informal systems and need quickly to develop something formal in their place. The mistake of assuming that everything is carrying on all right can be a dangerous one. The following examples show two contrasting schools, one shunned by nearby middle-class parents because the school draws heavily from the council estate on which it is based and the other in an area of low- to middle-income private housing and considered by parents to be a 'good school'.

Managing weak teaching

In one school a new teacher was having significant difficulties. He could not control the children. The work planned for them was not a good match to their needs, and the planning was weak. Activities were not well supervised. Colleagues offered some support, but management remained largely unaware of the scale of the problem and did little to support the new member of staff despite the significant investment of public funding in this individual by way of initial training. This failure to recognize the problem and provide adequate support meant that the problem got a lot worse before it got better. The class made no progress during a whole academic year and some children actually regressed in their literacy skills. The self-confidence of the new teacher was severely prejudiced by this experience, and he left the school with no job to go to.

In another school, three new teachers were being closely monitored by the headteacher, her deputy and other identified members of staff. Their planning was checked weekly. Their teaching was observed regularly. Their performance was discussed and support was offered. All of these teachers, to a greater or lesser degree, were experiencing problems, either related to controlling their class or to matching work accurately to pupils' abilities. In addition, some of their subject knowledge, in science for example, was weak. One of the teachers decided, after almost a year, that he was in the wrong job and left. The two others, both of whom had initially shown weaknesses similar to the new teacher in the first school, gradually grew in confidence and became valuable team members with successful careers in education ahead of them.

The first school in the above illustration was the school deemed to be successful in the local community, whilst the second was the school shunned by many residents in its own locality. Given the same material, two craftsmen of differing skill can produce artefacts of stunningly different quality. Likewise, given children from the same background, different schools with different value systems can produce academic results with these children of remarkably contrasting worth. Given newly trained teachers, willing but nervous, average in terms of their training and their own academic performance, two different management styles and two contrasting sets of professional expectations can produce either demoralized, poorly motivated individuals or 'gifted', energetic and valuable members of staff.

When an inner-city school is experiencing problems, it is easy to blame the teachers. Yet give many of those teachers decent management and they will perform at least as well they would in a more favourable environment, for a well-motivated teaching staff is a key feature in the success of any school. In such schools, teachers have a professional energy and the schools are vibrant and energetic places to work and learn in. Visitors can sense this factor. The school has an air of purpose because the management knows exactly where it is going.

Finding the right person for the job

The lesson of the previous paragraph does not mean that any teacher will suit any position in any school. Before appointing a new teacher, a school should analyse its needs carefully. If the special educational needs specialist has just left, there will probably be a case for replacing this teacher with one with similar skills. If a history co-ordinator has just left a small primary school, and has left the subject in good shape, there may be a case for seeing if a different set of skills, for example in art or geography, might be useful so as to develop another area of the curriculum.

Small schools particularly need to look carefully at the 'experience profile' of their staff. Young teachers need role models and need the benefit of more experienced colleagues to help them overcome some of the common difficulties and problems faced by all new entrants to the profession. Experience, though, costs money. It also requires the head-teacher and governors to define carefully a senior position, and this may be something that they themselves are not used to doing. It also requires a carefully thought-out job advertisement. There is the risk that an appointment may not be made.

Because of these factors, and the fact that many headteachers and governors can be reluctant to narrow down a potential field through stipulating particular requirements, the temptation to appoint from the local pool of newly qualified teachers is very strong. These teachers are new, fresh, cheap and are unlikely to challenge management. There is often a feeling that they can be 'moulded' to fit into a school's ethos. This is particularly the case in schools that suffer from weak and insecure management, who may see the appointment of more experienced and senior staff as a threat to their own perceived professional credibility. This strategy is lazy and unimaginative and should be resisted. Schools end up with no depth on the teaching staff; with no individual capable of running a major initiative, and with nobody who can really help the new teacher who is struggling when the going gets tough. Teachers leave these positions early – and sometimes leave the profession because of their negative experiences.

Every school needs a balance of older and newer staff. Every school can benefit from new ideas but must also benefit from experience and the wisdom acquired through experience. Often, a school is having difficulty appointing a teacher to a vacant post because a previous incumbent has left and there is insufficient time to recruit publicly before the new term. In these circumstances, it is better for the school to get by with the assistance of an experienced supply teacher until the right person with the right skills can be found – such an approach is certainly preferable to appointing anyone, particularly a newly qualified anyone, simply to get a body in front of the class. Ultimately, parents will place higher value on the school that insists on employing the best-quality and most appropriately skilled teachers rather than the school that simply fills gaps. Schools, and parents, need not fear good supply teachers. Career supply teachers will often do an excellent job in these circumstances, with their experience and adaptability being important assets for a school in a moment of particular need.

Appointing the right headteacher

School governors have a difficult job to do wherever they are in office. It is a fact that the job is often more difficult in an inner-city location. This is often because there are fewer people living in these areas who are willing, or perhaps able, to be governors. However, a small group of dedicated individuals can make a huge difference to the success of a school. The appointment of a new headteacher is one decision that many governors will have to make at some time, and it is arguably the most important decision a governing body ever has to make.

Imagine the case of a school that has been 'ticking along' for years. Everyone knows it could and should do better. However, in the absence of any major crises or significant failure of pupils to progress, there has been no real impetus to address the school's development rigorously. Now the head has gained promotion to a job in a larger school and the governors have their opportunity.

The first round of advertising produced little to inspire. Advertising is expensive and the school's location, in one of the tougher and less attractive parts of town, may be an inhibitor. Undeterred, the governors ask the deputy head, in post for just a couple of years and only of average capacity, to act as headteacher for the following term despite him not being seen as a serious contender for the position.

The second advertisement attracts little more interest, except that now the acting head has put in an application. The governors handle this situation tactfully, explaining that it has not been possible to draw up a sufficiently large field and so the post will be re-advertised. The deputy head is confirmed as acting head for the remainder of the academic year.

A third round of advertising produces no more than the previous two but leaves the governors with a real problem. Having had an acting headteacher for a year who has not done a bad job of managing the school and who has at least maintained the status quo, and having not attracted anyone of higher calibre, the governors wonder why they should not appoint the acting head. Although this candidate isn't the dynamo they wanted, the arguments in favour of appointment become irresistible. 'He knows the school well.' 'He understands our problems and needs.' In addition, 'What will it do for internal staff relationships if we do not appoint this person?' The acting headteacher is appointed notwithstanding that, as deputy, he was partly responsible for creating the current problems and not addressing the school's real needs. Such appointments certainly will not provide the vision that external impartiality can bring to a new position.

Several years down the line, the governors may well regret their decision. Target setting, heralded by the government as the strategy for keeping everyone on their toes, from pupils to chief education officers, is going to make sure that many governors do indeed regret weak decisions of the type described. Similarly, a governing body may appoint a knowingly weak candidate because they cannot see any other way forward, and this too is only trouble stored for the future.

So what went wrong in our example? Simply put, the school did not attract the candidates it wanted and had to settle for second best. It did not sell itself. Was it the advertisement then? Yes, in part, but not entirely. The failure to sell itself started a long time before the advertisement went to print. When it appeared in the educational press it read along the following lines:

St Optimist's VA Primary School
Gas Works Lane
Belfast-upon-Mersey
BM2 6GR

Dynamic headteacher required to lead this successful school, described by Ofsted as a caring and safe institution. The school is fortunate in having a committed teaching force and a dedicated governor and parent body.

Application forms available from the acting clerk to the governors c/o the school. Please enclose SAE.

The problem was that most of the other 124 advertisements for headteachers appearing that week in the educational press read along the same lines. The readers, aspirant headteachers with an eye to the move beyond this one, didn't deselect it solely because of its location but because it was just like all the others. At the end of the day, if there is nothing to make a job stand out then the location is as good a way as any of deselecting a school.

Building in strength from the top down

Aspirant headteachers of any quality want more than just a job. They want an opportunity and a challenge. If an advertisement doesn't show direction and describe a willingness to move forward then, in the eyes of its readers, the job may not have the opportunities to make a difference that a new headteacher may want. The advertisement can give readers an indication of the vision possessed by the governing body.

So what should the governors have done?. First of all they should have identified the key strengths of the school. It would have been useful to get the parents, teachers and pupils to help with this. These groups may well have identified things that governors take for granted or simply do not notice. The governors should also have identified the things that they wanted to change or develop, tangible things like:

- ❑ the acquisition of a sporting or musical reputation;
- ❑ the raising of standards overall;
- ❑ developing the school's facilities for community use;
- ❑ developing community education within the non-English-speaking section of the parent body;
- ❑ strengthening ties with higher education and developing a role in teacher education;
- ❑ developing pre-school education.

All this adds up to vision and, in the eyes of potential applicants, where there is vision there is opportunity.

What the governors need to do is to acknowledge the school's shortcomings, describe them as developmental points, and present their resolutions as aspirations. Then there will be a much better chance of attracting candidates who are in sympathy with the vision of the governing body than through a general run-of-the-mill advertisement. It isn't simply a question of attracting a field for a vacancy, but of attracting the right candidate to lead the school in the fulfilment of its aims. Careful analysis of the school's needs and imaginative communication of a vision for the future is going to increase significantly a governing body's chances of successfully making a good appointment. It may still take three advertisements, but the governors should eventually get the candidates who fit the bill and not have to settle for second best.

One final point. Prospective candidates are put off by a school that appears to be so short of funds that it cannot even afford the postage stamps to send out information packs. If a school is expecting several hundred enquiries, then perhaps there is justification in asking candidates to enclose SAE (stamped, self-addressed envelope); however, if it is only expecting, say 30 or less enquiries, why risk putting off potentially good candidates by trying to save a few pence? False economies such as this can prove very costly in the longer term.

The revised advert could then look like this:

High-performing individuals who can make a difference are invited to apply for the position of Headteacher of St Optimist's VA Primary School.

We are already recognized for:

- ❑ the high quality of care that we provide for our children;
- ❑ the commitment of our teachers;
- ❑ the dedication of our volunteer governors and parents.

We now need an able individual who can:

- ❑ raise academic standards above the local and national averages;
- ❑ build on our sporting and musical achievements;
- ❑ develop facilities for pre- and after-school care as a service to the community;
- ❑ make St Optimist's a school of which parents can be proud.

Application forms are available from the Chair of Governors, St Optimist's VA Primary School, Gas Works Lane, Belfast-upon-Mersey, BM2 6GR or telephone 01016–454545.

As a footnote to this example, it should be noted that it is very important that the governors are open and fair with the deputy head teacher/acting head from the beginning. If the deputy headteacher hasn't got the experience to do the job permanently, then the governors should say so. A clearly defined vision for the future will help governors in saying this. It may well be that they wish to insist on a given number of years' previous experience as a deputy headteacher, or even as a headteacher. Making this clear at the outset allows the deputy to gain valuable acting-headship experience without feeling offended or overlooked at the end of the process when an outsider with the required experience is appointed.

Summary

'Failing' teachers can be found in any type of school. Teachers who have been successful in one school may not be successful in another. Schools can change when the headteacher changes, and so the quality of teaching may begin to change also in these circumstances. Appointing a balance of younger and more experienced staff is essential. Equally essential is the appointment of the right headteacher and not, under any circumstances, settling for second best.

MYTH 2: 'INNER-CITY CHILDREN ARE UNTEACHABLE'

It is a commonly held opinion that the most difficult pupils are found in the schools with the toughest reputations and that these are usually found in the inner city. However, problems can arise in any institution where management has not recognized the need for structures and security when dealing with young people.

The nightmare scenario

Many people will remember the images broadcast on television in 1997 of the Ridings School in Yorkshire, where the pupils had become virtually uncontrollable. Cameras showed children walking up stairs behind adults waving fingers in offensive gesticulation. It would be reassuring to think that the children only behaved like that because the cameras were there and that this school was a truly remarkable and unfortunate exception in our education system. Unfortunately, other schools have had similar problems and other schools may well yet develop similar problems.

> In one large secondary school visited by inspectors in the mid-1990s, some staff admitted being nervous about walking down the corridor or staircases. The corridors and stairs in this school were exceptionally wide and large numbers of children could, and did,

move through them very quickly. An 'incident' could occur just a few feet away, but the perpetrators could be lost in a crowd very quickly. At its worst, the staff response to these problems was not to police these areas. When staff did attempt to control pupils' movement, their attempts were often isolated, not followed up, and not repeated consistently; they were seen by the pupils as being half-hearted. Unsurprisingly, teachers in some classrooms also felt threatened. The absence of security meant that in some practical lessons teachers were unwilling to risk using valuable, or potentially dangerous, resources. Consequently, the quality of the children's education was impaired.

In any school, in the inner city or not, a lack of systems, a lack of institutional expectations and a lack of orderliness will lead eventually to a form of institutional anarchy setting in. A school has a larger concentration of individuals permanently housed in it than almost any other working environment. In the absence of established systems, it is in the nature of human beings to organize themselves and to form groups to exercise influence or to take power. A school without a strong organizational ethos – without clear parameters for social interaction and without a clear work ethic – is leaving itself wide open to the formulation of trends, practices and relationships that will fill the vacuum left by the absence of formal control and organizational structures.

Many would consider that problems such as these are confined to secondary schools. It is true that secondary schools have, in the past, been the schools most at risk of behavioural breakdown. However, there is an increasing tendency for these problems to occur in primary schools as children, from all types of background, arrive with fewer and fewer social skills and with increasingly blurred perceptions of the difference between right and wrong. In both primary and secondary schools it can be possible for significantly serious problems to be occurring in one or two classes and for these problems to be largely unchallenged by a weak management.

In a primary school, the level of noise in one classroom was intolerable. It was not the busy noise that comes from children working together who sometimes get excited but generally stay focused. This noise was from groups of unsupervised children who had been left largely to their own devices while the teacher concentrated her efforts on a complex practical task with one group.

The children had been divided into groups and given independent, 'low maintenance' tasks to complete whilst the teacher could 'teach' a difficult skill to one group. The planning showed how it should have worked, but in practice it was a poorly thought-out strategy that left the majority of the class with unstimulating and unfocused written work to undertake. These pupils did not see the work as being valuable, and this judgement was reinforced in their minds by the teacher's own lack of interest in it, as demonstrated through all her time being spent with one other group on different work. The teacher failed thereby to manage the whole class.

When boredom set in for those not in the special group, noise levels began to rise and then one or two pupils left their seats to 'talk' to others. The teacher did not respond. Then paper missiles and the odd pencil began to fly, until eventually a fight broke out when a bit of pushing and shoving got out of hand. Then the teacher turned round, sent children back to their seats with a minimum of admonishment, and continued her work with the special group.

Nothing had changed in terms of the overall balance of teacher attention in the room or in the nature of the work given to the class, and so the cycle started all over again . . .

The school in which the above incident occurred was a suburban primary school, in a not particularly well-off, but certainly not underprivileged, area.

In both primary and secondary schools, the requirements for strong management, clear parameters for social exchanges, and a strong work ethic are important. Children in inner-city schools respond well to a properly thought-through educational experience in an environment in which they know what is expected of them, just as much as children in any other school. Children in the suburbs can respond just as negatively to a poor experience and will just as effectively find ways of manipulating a system that is weak in its requirements for consistent standards of behaviour and work.

Summary

When children rebel in a school it is not because they would have done that anyway. Children respond, like adults, to the circumstances around them. If no rules exist, then the pupils will invent them. If they perceive a weakness, then they will exploit it.

It is possible for single classes in small schools to behave in this way, and the effects for these children are just as serious as for those in the widely publicized cases where whole schools appear uncontrollable.

MYTH 3: 'OUTSIDERS DON'T UNDERSTAND THE PROBLEMS WE FACE'

There are teachers in inner-city schools who cast envious glances at their colleagues working in more privileged surroundings. The advantages that these fortunate individuals have seem to be endless. They can spend their days teaching and not doing 'social work'. Children come to school having had breakfast and arrive willing to learn. They return their homework, completed and ready to mark. A school can lend them books and resources that are essential to their learning and knows that it will get them back. The pupils concentrate in lessons and respond to questions. There is less stress in the classroom and less stress in the staff room – there has to be, simply because everything is working as it should be.

What fortune is it that sees some teachers working in these desirable circumstances whilst others have to tough it out in the inner city? Whatever it is, talk to teachers in the more 'privileged' schools and a different picture emerges.

Understanding deprivation

A teacher moved from a most deprived inner-city environment to a 'privileged' and comfortable suburban setting. After she had spent 14 years in the inner city, the move was a major culture shock for her. There were some remarkable differences and, true, some of her new colleagues didn't have a clue about many of the issues that she had become all too familiar with in the inner city – for example, the number of children with English as a first language or the number of children from very low-income and often unstable households.

Here at her new school, children arrived in large cars driven by nannies and free school meals were almost non-existent. The head-teacher, who had also taught in the inner city, cautioned the new staff member about only looking at the surface. She would say that deprivation here was a six-year-old having five au pairs before Christmas. Large numbers of these children saw little of their parents. They had everything that money could buy but very little of their parents' time and attention, and this neglect led to its own problems – problems that would be familiar to teachers of 'disturbed' children in the inner city. There may have fewer 'disruptive' children, overall, in the classrooms of this school but the challenges faced by their teachers were no less demanding.

Under-privilege is comparative and that there are different types of deprivation.

Different schools and different challenges

One feature of middle-class schools is the extent to which parents make heavy demands of individual teachers and headteachers to try to get a situation resolved the way they want it resolved. Now, there is nothing wrong with parent power; and there is nothing wrong with trying to get the best out of a system for your own child. There is certainly nothing wrong with challenging a school if it is thought that the school is giving a poor-quality service.

What, though, about the parents who persistently demand additional homework for their children, despite the homework being set in line with the school's well-publicized policy for homework? Or the parents who upset their children almost nightly by telling them that their homework is 'rubbish' and by writing negative comments all over it? Or the group of parents who disagree with the way a teacher has dealt with one child (usually somebody else's) in their children's class and who are demanding that the child be punished again? Or the group of parents who all threaten to withdraw their children if a child with a particular special need is not excluded from the class or given special tuition? These are not isolated incidents. They occur daily and weekly in classes and schools throughout the country. They put just as much pressure on the teachers in these schools as persistently disruptive pupils do in other schools.

It is also inaccurate to believe that only teachers in inner-city schools are at risk of being subject to violent attacks. The threat of physical violence is not by any means confined to the inner city. Teenagers anywhere – boys *and girls* – will, if they think the occasion warrants it, resort to violence in the classroom or the playground. Local casualty units can testify to the viciousness of some of these incidents in seemingly middle-class, secure institutions.

There is really no such thing as a 'typical school' and no such reality as 'a perfect school'. Every school has its own very specific and very unique set of challenges to face. There was a time, perhaps, when teachers in more 'privileged' schools could claim success on the back of an already able intake. However, increasingly sophisticated systems for measuring added value mean that the pressure on teachers in these schools to make a difference is just the same as it is in any other school.

Every school serves a unique community. Within that community there is a range of contrasting social and economic factors at play; there may be significant environmental factors too. The buildings and school grounds play their part in shaping the nature of the school community – for example, they dictate the range and frequency of sporting activities or whether pupils can independently be left in certain areas at break or at lunchtimes. Then there is the headteacher and the teaching staff, particularly the headteacher. How committed is he or she? What vision does he or she have? How effectively are the teaching staff led? How well is the school's image projected within the community? How experienced are the teachers? Do they have the right range of skills for the school? Are they well trained?

In the final analysis these issues will be amongst the factors that will decide whether a school is successful or not. Children everywhere have a right to be educated properly. Teachers everywhere have a right to be well managed. There is no excuse for governors settling for second best when it comes to appointing – or retaining – their headteacher. And, in turn, there is no excuse for headteachers settling for second best when it comes to appointing, or keeping, their teachers.

All the problems that a school faces – and different schools face different problems – are part of the professional challenge that must be overcome in order to do the job properly. They are not, and cannot be, excuses for failure.

Summary

Every school is different. Because a school is in an affluent suburb, it does not mean that there will not be some unpleasant challenges for the teachers. Because another school is in a tough inner-city area, it does not mean that it has a monopoly on deprivation and disturbed children or on abusive parents. Each community presents a particular set of circumstances and challenges to teachers, headteachers and governors of the schools it feeds.

CHAPTER SUMMARY

The myths described in the preceding sections are ones that apply particularly to inner-city schools, but they are indeed myths that need exploding. Experience shows that schools of all types and all locations can face difficulties, and the main points arising from this chapter are summarized below:

- ❑ The excuses used by people defending poor performance in inner-city schools are excuses for unimaginative management, low expectations and lack of vision.
- ❑ Poorly motivated teachers are found working in many schools, whether in the inner city or not.
- ❑ Average teachers with poor leadership and no support can soon become demoralized, with a subsequent effect on the quality of their work. Supporting and motivating staff to teach well requires credible and effective management and not the luck of having a 'gifted teaching force' (as is often heard in weakly led schools).
- ❑ It can be difficult to recruit headteachers and teachers to inner-city schools – but it need not always be so. Governors must avoid appointing a knowingly weak candidate into headship because that candidate was 'our only choice' or 'the best of a very poor field'.
- ❑ Similarly, headteachers and governors must avoid appointing classroom teachers who do not fit their recruitment profiles and who have dubious previous experience 'because we need the body in front of the class' rather than because they are what the school wants or needs. Such strategies actually compound a school's difficulties rather than helping resolve them.
- ❑ Headteachers and governors will often sacrifice long-term stability, growth and success by appointing, on a permanent contract, a teacher who is unsuitable for the position. This then results in additional obstacles to school improvement and even perhaps expensive and time-consuming disciplinary action in the future.
- ❑ Schools anywhere – but especially in the inner city – are more likely to fail because of weak governance and management resulting in weak teaching rather than because they have inherently weak teachers or 'difficult' pupils.
- ❑ Every school has to work with the problems of the community it serves, and children's progress can be affected by these problems. These problems are aspects of a school that require effective management. They are not excuses for failure.

3

RECOGNIZING THE SYMPTOMS OF FAILURE (1): TEACHERS

Having examined other outward signs of inward failure at institutional level, it is now time to look in some detail at the qualities of a school's greatest asset – its teachers. If the teaching force is not operating at an optimum level, then the performance of the school as a whole will be depressed. Every aspect of teaching is considered in this chapter, from the initial preparation for teaching, including the use of assessment to guide planning for individuals or small groups of children within a class and the choice of teaching techniques, through to the organization of the classroom, the management of classes and the presentation of lessons.

PLANNING

So what are the attributes that make a good teacher? What is it in some teachers' work that makes them more successful than their fellow graduates? What is it that makes the staff of one school gel more closely together as a team and makes them more effective in their work than the staff of another school? The first consideration is how the staff, as a team, plan their work and how well individuals within that team follow the agreed approach.

Planning for perfection and consistency

The first place to look when assessing the quality of a school's teaching provision is the quality of its planning. New teachers almost always plan beautifully. Often they overplan – at least for the first few weeks of their new career! If they are unlucky enough to be employed in a school where teachers' planning is 'informal' and not monitored properly by management, they will soon get wise to the fact that they are the only ones putting so much time and effort into their planning. The smart ones will realize quickly that that is why their lessons actually work, and they will appreciate that a more 'informal' approach to planning may be one of the rewards of experience that could be reaped *later on* (but not now).

The really smart ones will realize that only by careful planning can they properly meet the different needs of all the children in their class. However, it is difficult for new teachers to sustain the best elements of their training if it is the case that, when they get into the real job and are working alongside real professionals, the majority of other teachers have apparently abandoned these principles.

Sometimes a visitor to a primary school, or to a subject department in a secondary school, will see everything appearing to be working in harmony. The atmosphere is the same in every classroom; there is an air of industry and purpose in all the classes; all the teachers appear confident in their work; there is little evidence of short-temperedness or shouting by teachers; the children appear happy and settled. In contrast, another school, or another subject area in a secondary school, can feature one or two classes where the atmosphere is different: where pupils are making more – or sometimes less – noise than those in other classes; where some teachers appear to shout more than others; and where some teachers struggle with the curriculum they are teaching. All children have 'favourite' teachers, but in this school some years and some classes are very unpopular with parents as well as with children. It is almost a certainty that this latter category of school, or the teaching teams within them, will have very different planning methodologies from the former category, and that these different approaches contribute to the atmosphere and ethos within each school.

The school with a standardized system for planning will appear to be the more cohesive institution. This is the school where the expectations of each teacher by management in terms of planning are consistent, and where the planning methodology has been carefully thought out so that it serves the needs of the school. This is the school where teams of teachers plan consistently at a high level in order to support a shared understanding of what they are trying to achieve. Planning is considered to be a vital part of the teaching process. Time has been invested in identifying the right format for planning and in identifying the information that teachers need to record for their planning to be useful.

Long-, medium- and short-term planning

Often, schools will invest a great deal of time in devising long- and medium-term planning strategies. These define what it is that each teacher will teach during the course of a term or a half term; together, they provide a detailed curriculum framework that can be mapped and matched against the National Curriculum or other required syllabus.

It is in the area of *short-term* planning that practice varies too much from school to school, and sometimes from teacher to teacher. Short-term planning is that which is done weekly or daily. It usually refers to the subject content, but is more concerned with the logistics of the day and with interpretation of the subject content for pupils in different ability groups. When you see this kind of planning, which does not duplicate or repeat the long-term planning, you know that you are looking at the work of a teacher who understands that the learning needs of her pupils vary from group to group, from individual to individual, and from week to week.

Table 3.1 sets out the main points relating to each of the three types of planning that should be used in schools.

Teaching without planning

So what happens when a school does not have any formalized planning systems? Firstly, the teachers are all working in a vacuum. Rather than be

Table 3.1 *Planning structures*

TYPE OF PLANNING	CONTENT
Long-term	Relates directly to schemes of work Summarizes the broad areas of learning
Medium-term	Relates directly to long-term planning Includes broad Learning Outcomes and assessment criteria Includes major resourcing issues (use of non-routine accommodation, equipment trips or visiting specialists) Indicates differentiation in curriculum content and pupil outcomes
Short-term	Relates directly to medium-term planning Includes specific Learning Outcomes and assessment criteria for all ability levels Includes assessment methodology Shows differentiation strategies Shows organizational and teaching strategies Shows resourcing requirements Shows lesson structure Shows homework opportunities

guided by an overall picture that indicates roughly what content should be covered in each year, the teachers are left to decide for themselves what should be taught to their classes.

Despite a rigid structure and a high level of prescription, the National Curriculum does not guard against the worst excesses of this problem. The National Curriculum indicates what children should know, should be able to do and should understand at certain stages in their education. However, children are different, and they progress at different rates. A strictly interpreted National Curriculum can be as restrictive to some children because it holds them back as it is inaccessible to others because it is beyond them. It will be interesting to see whether the much-vaunted Literacy Hour and Numeracy Hour will be of real value in schools where there is already high attainment in these areas: some schools really do run the risk of joining a (non-statutory, by the way) band-wagon and sacrificing their current good practice because parents and governors expect them to be at the forefront of the latest educational initiative.

Secondly, in the schools where there is no whole-school long-term planning, there is a danger that chunks of the curriculum may be missed out or that some sections may be covered twice. Meeting children's needs becomes a hit-and-miss affair because teachers are uncertain as to what has been taught before. In the worst instances, children can arrive in a SAT (Statutory Assessment Test) year and be unprepared for their assessments. This is the point when the problems become so obvious that action has to be taken. The scheme of work needs examining, revising and breaking down to match the needs of the pupils in each year.

So what happens when a school has the long-term planning in place but doesn't insist on a corporate style for short-term planning? Well, sometimes it works and sometimes it doesn't. In the best instances, the quality of the long-term planning is such that, together with a good fit of the lesson content to the pupils' abilities, it informs the practice of individual teachers well. These instances are more likely to occur in schools where there is an experienced staff with an element of team teaching that helps moderate the curriculum in each classroom. There are risks, even so, that new teachers may develop planning and teaching habits that go undetected for too long before being picked up by colleagues or management. There are also risks that the needs of individual children may be overlooked because of the more general perspective that the lack of detailed short-term planning leads to. In the worst cases, teachers adopt a broad-brush approach that neither stretches the more able nor moves forward the less able in their classes.

If the quality of medium-term and long-term planning is weak, it may not provide a sufficiently informative framework for formal planning, let alone informal planning.

Streaming or setting does not necessarily mean less planning

One of the perceived advantages to teachers of focused-ability teaching strategies, as opposed to mixed-ability teaching, is that planning should be easier. This is because of the narrower range of abilities found in each set, group or class. If children's interests are being truly addressed, though, these strategies should lead to more focused teaching for *each individual child* rather than a blanket approach for the whole class. The true advantage of having a class or group with children of a similar ability is that the difficulty of the work, whilst broadly the same for each pupil, can be adjusted to optimize the learning potential of each child. This is where real advances in the rate of learning can be made.

Setting and streaming give best results where there are opportunities for pupils to move to more challenging contexts or to situations where they may receive more help. There are several problems with this that need addressing in order for the system to work to the pupils' best advantage. These are as follows:

- ❑ A high level of knowledge of pupils' specific abilities, strengths and weaknesses must be known in order to inform the decision making process.
- ❑ A school needs to have the mechanisms in place to regularly review progress.
- ❑ Schools need to have the flexibility to be able to move pupils between groups and classes without overloading any groups or classes.
- ❑ Where pupils are moved 'down' to a group doing lower-level work, this arrangement needs managing sensitively with pupils and their parents. The schools strategy could be described to parents as a move to a context where they will receive more specific support to address identified problems rather than a move to a 'lower' group or set.

All these issues imply a high level of planning, informed by accurate assessments, so as to ensure that the needs of individual children are properly targeted.

The Learning Outcome

All planning, at whatever level, is enhanced through proper understanding and use of the 'Learning Outcome', sometimes called the 'Learning Objective'. A well-defined learning outcome will:

- ❑ articulate exactly the main element of knowledge, skill or attitude to be acquired in the lesson;
- ❑ be very measurable, given that it may be either achieved or not achieved by the children;
- ❑ be a marker that shows children and their teacher when a particular point in the lesson has been passed.

Well-defined Learning Outcomes are nothing short of essential ingredients in good short-term planning. They can be different for different groups of pupils. This is the point when general lesson content takes on a particular focus for different children depending on their previous knowledge and on their abilities.

The absence of Learning Outcomes leaves a space in a teacher's planning. It is as if that teacher's intentions for a lesson are vague and ill-defined. Sometimes a lesson doesn't work and teacher and pupils get to the end of it wondering what went wrong. These lessons are often the ones where the teacher has not decided the Learning Outcomes that apply. If the teacher's own aims are unclear, don't be surprised if the pupils' learning is also fuzzy as a consequence.

Good short-term planning isn't only a record of what has been or will be taught. It is a system of goalposts and markers by which the teacher can measure the success of the operation. It needs to be not vague but specific, as illustrated in Table 3.2.

Planning and target setting

Schools are moving into an era of target setting. Target setting is a strategy that encourages whole-school improvement through concentrating on the attainments of certain groups within the school community. Target setting requires teachers to know precisely the abilities of each individual within the classroom, and with this knowledge the next stage of work can be carefully chosen so that children move one step closer towards achieving their particular individual or group targets. This kind of teaching relies heavily on the teacher being able to plan the work of individuals and groups in the short-term, so as to maximize their chances of achieving set targets.

Table 3.2 *How to express Learning Outcomes*

CATEGORY	EXAMPLE LEARNING OUTCOMES
Good Learning Outcomes	Children are able to use the library, class reference books and the computer to find information on Henry VIII. Children are able to add and subtract using numbers to 99. Children are able to demonstrate knowledge of the correct use of verbs and adjectives in their creative writing. Children are able to demonstrate understanding, through their recorded accounts of experiments, of the different properties of given materials. Children are able to demonstrate knowledge, through their written work and successful completion of homework tasks, of health conditions and services in Victorian England.
Poor Learning Outcomes	Start project on Henry VIII. Complete mathematics worksheets on addition and subtraction. Write verbs and adjectives. Carry out experiments using water and weights on different types of paper and fabric. Learn about health conditions in Victorian England.

Target setting is too valuable an initiative for the whole school to be left to the informal planning strategies of individual teachers. At a time when teachers are becoming collectively responsible for all pupils' attainments within their school, they have a right to expect corporate systems that will help make the job easier and that will help ensure consistency in the quality of practice throughout the school.

There is one other good reason for short-term planning strategies to be consistently deployed across a school. Often, supply teachers or internal cover teachers arrive at a classroom to find no indication of what the children have been doing that week and what the next stage of their teaching and learning should include. Copies of the previous day's or week's plans would help enormously. It is also curious that teachers and schools have been so remarkably slow to see how information technology can help them with their short-term planning. Perhaps it is simply a case of teachers being too busy to take on board another new initiative, particularly when it requires them to learn new skills. However, as technology becomes easier to use and school networks become more common, school managers and teachers should consider carefully the time that could be saved in the long term through a short-term investment of time put into developing new skills that would enable them to plan effectively using information technology.

Summary

The presence of an institutional system for short-, medium- and long-term planning will help ensure consistency of teaching throughout a school. It makes it easier for teachers to match their work precisely to the needs of children of differing abilities. The use of good-quality Learning Outcomes helps to focus the teaching content and is an aid in assessment. Structured planning helps a school to focus on its targets at pupil level.

PRESENTATION

How teachers face their class is paramount to the effectiveness of their teaching. In the same way that the first few minutes are important to establishing an effective relationship at adult level, the initial contact with a class – and, to a similar extent, the opening of every lesson that follows – directly influences the outcome of the lesson. The way that the teachers talk, the way that they present the lesson content, the chosen style of teaching, and the quality of their own resources and board work are all factors that can influence the success or failure of a lesson.

'Simon says'

Children learn by example. They learn by imitating adults and by imitating each other. Children like role models. Any adult in a school is a role model and will influence some or all of the children in a lesser or greater way. The single most influential individual at any one time, outside family, in a child's primary school will probably be the child's teacher. When a teacher has a strong and positive personality, it is easy to see the influence – even at secondary school level.

> The head girl of one school, at the end of the school's annual play, publicly thanked the drama teacher for her work in putting together a particularly good production. The head girl had all the drama teacher's mannerisms, nuances and even sounded like her. Uncanny!

Primary children tend not to mimic outwardly so obviously, but they do instil in their own behaviour and attitudes many of the attitudes and attributes of their teacher. How many parents have argued with an eight-year-old to undertake a task in a particular way, only to be told by the child that the procedure described is wrong and there is only one way to do it 'because Miss Smith says so!' There really is no answer to that most of the time!

When a teacher has a weak or negative personality, the influence that this can have on the children in the class is still evident but is altogether more damaging. What is more, some parents may not link changes in their children's behaviour with exposure to negative attitudes in school. Take the newly appointed teacher who wants to make a good impression in a new school. The teacher knows that initially it is wise to adopt a quite strict, no-nonsense approach with a new class until pupils and teacher get to know each other. Mistaking aggression or oppression for strictness is a relatively common occurrence for some new teachers. Eight-year-olds, used to a firm but gentle style of classroom management, may become frightened if this is replaced by what they perceive as real anger and real threats from an individual that they do not know. For the more nervous, and for some of those children who, one way or another, attract this type of attention from the teacher, this can lead to a loss of concentration and a loss of enjoyment in their work. At home they may suffer disturbed sleep and a state of general upset, resulting in a reluctance to go to school. Other children may learn from the teacher's role model that it is good to be aggressive and to get angry because that is how you get others to do what you want.

Do as I say, not as I do

It is not only through general demeanour or overall personal style that a teacher influences children. The quality of a teacher's own work acts as a benchmark for the children in the class. Teachers who have themselves developed a tidy and well-ordered style of board work will have less difficulty in persuading their pupils to be careful in presenting their own written work than teachers whose style is untidy and haphazard. In contrast, it is very difficult for teachers to find fault with children's uneven writing when their board work slants 160 degrees to the top right-hand corner!

Teachers who produce children with the best handwriting generally have, in the classroom at least, excellent handwriting themselves. When you are teaching young children to form letters properly, it is important that every occurrence of that letter that these children see in the classroom, particularly if produced by the teacher, is formed properly.

Where a school is having difficulty teaching and promoting good handwriting, even in accordance with its own policy and adopted handwriting scheme, it is often because some teachers are not familiar or confident enough in the handwriting scheme themselves. This results in the pupils being taught an inconsistent handwriting style or, in some classes, not being taught at all, as they move up through the school. Imperfections and bad habits that appear after children have been initially taught to write can be difficult for teachers to address, particularly if the teachers for the older classes are not handwriting experts themselves. These difficulties are also compounded in older years because of the pressures of the curriculum; less time is available for handwriting exercises. As a consequence, one of a child's most personal attributes, against which a judgement will be made many times in the child's life, may already be being compromised and sacrificed before the child has even left primary school.

Death by worksheet

One of the worst examples of poor presentation by teachers – all too commonly seen – is the much-maligned (but often rightly so) 'worksheet'. Reprographics have come a long way since the blue-ink Banda copies of the sixties and seventies. Unfortunately, many teachers' skills in preparing their work for copying on the newer technology do not do the higher quality of the reproduction techniques any justice at all. We are not necessarily talking about using word processors or computer presentation packages here. Whilst these can be very attractive and do make, in the hands of the practised user, a very good visual impression, attractive and useful results can be obtained from handwritten resources. This is particularly the case if one of the purposes of the worksheet is to promote handwriting skills.

One problem often seen, particularly with mathematics worksheets, is too much information and too many questions on the page. Such sheets look boring, spacing imperfections become exaggerated so the sheets look messy, and the whole thing is a turn-off to otherwise motivated pupils. Furthermore, it is overwhelming to those pupils whose mathematical knowledge is less secure. Another common problem is the inadequate worksheet – one that contains closed or irrelevant questions that the children can complete even before the teacher has finished giving it out!

Sometimes questions, particularly in topic work, can be obscure, leading to a stream of requests to the teacher for assistance – and thus undermining the purpose of the worksheet, which was to encourage independence in the pupils' learning. All too often all pupils get the same worksheet or there are insufficient variations in the worksheet for groups of pupils with different abilities. This occurrence is one of the single most common reasons for children slipping 'off task' and becoming distracted. If they cannot understand the language used on the worksheet, then they are effectively being excluded from participating in the lesson, often with predictable consequences for everybody else's learning.

The preparation of the worksheet needs to follow the same principles used in preparing the overall lesson plan. These are:

- ❑ What is it that the teacher wants each group of children to learn by completing the exercises on their sheet?
- ❑ Can the children's learning be evidenced by examining what they have written on their sheet?
- ❑ Will completion of the worksheet demonstrate to the child and teacher that learning has taken place?

Marking (What you sow, you reap)

Children look forward to having their work marked. For young primary school children, teddy bear stickers or smiley faces are recognizable indications that they have done good work. They are a valued sign of approval. In secondary schools a well-written comment giving credit where it is due and constructive advice when necessary can be as valuable to a pupil as the experience of the lesson itself.

Why, then, do so many teachers pay scant regard to what their marking looks like in a child's book? If comments are worth writing against the work of younger children, the writing should be clear and legible and the language used should be easily understood by the child. The only exception to this is where the teacher may use the practice of writing assessment notes directly onto the work of younger children. If this practice is being used, the use of a different coloured ink for notes intended for the child and those intended for the teacher's future reference would be appropriate.

Illegible writing or unclear comments are some of the most frequent occurrences in poor marking. As far as the affected children are concerned, they know that there is a problem with their work – hence in most cases the comment. The fact that the children cannot identify what the problem is from the teacher's own scrawl does nothing for their confidence and does nothing to help resolve the problems. For most pupils, having to ask the teacher for an additional explanation when they got the work wrong anyway is an experience that they would rather forgo – and they often do. The result is that teachers do not realize that their comments have not been helpful and don't appreciate that the problems still exist.

If teachers want to test this theory, they should go to a colleague's classroom and ask pupils what the teacher's marks on their work mean. Ask them if they understood why they got the work wrong (if it was wrong), and ask them what their teacher expects them to do about it. The answers may be surprising! Parents can, and must, ensure that teachers' comments on their children's homework are understood by the child and acted upon.

Classroom or shop window?

The other aspect of a teacher's presentation that influences their pupils is the quality of the display in the classroom. This does not refer to the abilities of teachers to use their own art skills artificially to improve children's work (despite this practice being commonplace in some schools); nor does it refer to those schools or classrooms whose use of display turns the place into a riot of colour and airborne obstacles, where it is difficult to walk without bumping your head on a 'witch on a broomstick' or a 'firework' or perhaps even a painted paper plate – in such rooms the display would probably be condemned by the local fire officer! Rather, this aspect of presentation concerns the ability of teachers to mount and label pupils' work in a way that does credit to the children's efforts and gives them real pride and a sense of achievement in their work.

Schools increasingly have teachers in charge of displays in public areas, and there is usually somebody who knows all the tricks about double mounting and the use of colour and texture in display who can be persuaded to share their skills with less 'talented' colleagues. It is surprising how many schools still do not realize the importance of that first impression given to prospective parents or other visitors through tidy, well-presented displays. One of the most easily spotted errors is, surprisingly, spelling mistakes in the labelling of pupils' work. Now, here we have a problem. If a pupil has labelled the work, then some teachers would not want to correct it because it takes ownership of the work away from the pupil. But surely a teacher's responsibility is to all the pupils who may use the display as an informative reference point for their own work, as well as to the individuals' own development and learning. Whereas there may be some justifiable arguments for leaving uncorrected mistakes in the body of the text of some pupils' work, there is no excuse, in the primary classroom, for spelling mistakes in the labelling of pupils' work. The use of a dictionary should be an automatic part of the labelling process.

> An inspector visited a school and in one classroom saw a display of words forming a vocabulary aid for pupils. The word 'Telephone' was mis-spelt as 'Telefone'. The very young teacher in question had not realized her mistake and appeared clearly embarrassed when the error was sensitively pointed out to her at the end of the lesson. End of story – or so the inspector thought. Unfortunately, a repeat visit to this classroom several months later showed the offending word card still to be on display. This time it was mentioned to the headteacher. Unbelievably, several months later it was still there!

Notwithstanding that the whole episode may have been an elaborate ruse to test the inspector's patience and consistency of judgement, it is alarming that a teacher could have so little pride in her own work. It is equally troubling that a headteacher could ignore the mistake for so long, and that both could be so insensitive to the learning habits of young children by leaving this error on display for such a lengthy period of time.

Secondary schools are often the worst offenders when it comes to poor-quality displays. In many instances, when you leave the front entrance and the art department behind, most of the rest of the school is a barren desert. It is often said that the reason for this is that pupils' work would be vandalized if put on display. There are two kinds of damage that can occur to display material in secondary schools. The first is caused by the person who mounted the work in the first place – for example by using an insecure mounting or by placing it on a wall used by pupils to lean on when awaiting access to a classroom; don't be surprised in these instances if the work becomes detached and develops a tatty, dog-eared look. The second is deliberate vandalism by pupils; however, part of the education process is teaching children to value and respect the property of others. Presented properly, even in the toughest of environments, vandalism need not be a particular problem to displays in the public areas of schools.

Take, for example, a very tough inner-city boys' school in a northern city, as described in the next case study.

The school, now closed, was sited in a very poor area and was populated mainly by white Roman Catholic children of Irish descent. The outside of the school was a shambles. Built in the 1960s of mainly glass and concrete, at times there was hardly a Monday morning when a ground-floor window was still intact. The staff room was depressing: no natural light, and just hardboard panels until the governors could afford another visit from the glazier. Many classrooms were the same. The vandalism though was caused mainly by local youths who attended other schools. Yet the inside of the school was very different.

The deputy headteacher, having retired from his successful business as a sign writer and having found rapid promotion in his new career in teaching, believed that if he gave the boys something to be proud of in the building, then they would play their part in taking care of it. So the corridor outside the history rooms developed cobbles (real ones) and the doors and walls developed beams and period surfaces. The geography rooms developed grass-hut style doors and walls, and large artefacts – ranging from suits of armour to crocodile skins – appeared in the respective corridors. That was not all. The local arts and libraries department must have been delighted to receive a request from, by repute, an uncultured school in an uncultured area, for the long-term loan of 50 or so framed pictures. So the Constable landscapes and Turner seascapes arrived, dusted down from the deep storage that they had occupied for several years.

They were, though, never displayed as they were originally intended to be. Each one was painted over and used for attractively framed charts and tables of figures showing the progress of anything and everything that could be measured and encouraged in this deprived school. School football results and other sports results and achievements were recorded. Inter-house contests were reported, and house point scoreboards were updated. Music and drama information was displayed. All these boards were updated almost weekly. None was ever vandalized, and pupils took real pride in the internal appearance of their school, particularly as it demonstrated their achievements and was relevant and, in the case of the history and geography areas for example, interesting to them.

The children never knew the source of the framed displays they enjoyed so much, and so the potentially dubious morality of painting over these loaned pictures was lost on them! However, it could be, and indeed was, argued that these pictures, in their new incarnation, played a much more valuable role in educating, motivating and inspiring those particular youngsters than they ever would have done in their original form. It is interesting to note that the incidences of vandalism on the outside of the building dropped as the inside of the building underwent this development.

In similar fashion a primary school in a particularly rough area in London has adopted this approach and prides itself on being able to display the most delicate artefacts within reach of almost all the children, without loss or damage.

Whilst these are two extreme examples of teachers using their initiative to make the school curriculum come to life and demonstrating trust to their pupils, there is no denying that the strategy worked. Make a public display interesting and relevant, and make it deliberately overt, so that you are showing the pupils that you trust them, and they will respect it.

Summary

Children learn by imitation. If teachers do not themselves demonstrate the standards they are expecting of their pupils, then at best the outcome will be diluted and at worst confused. Children deserve the highest quality of resources and worksheets. Work should be marked with a view to helping each child to progress. The environment in which a child learns can help the child develop responsible attitudes; it can support learning and encourage pride in well-produced work.

CLASS CONTROL

Whilst the best planning in the world will help significantly in making a lesson a success, unless the teacher can learn – and learn quite quickly – how to control a difficult class, the rest of the teacher's work will be undermined. Control isn't only about good discipline; it is about choosing the right teaching techniques and managing the atmosphere in the classroom.

The real super-teacher

Without doubt, one of the hallmarks of a really good teacher is the ability to control the class and to maintain discipline in such a way that the learning process continues unimpeded in the classroom. Those who demonstrate such skills are the teachers whose classrooms during the school day are pervaded by an air of industrious concentration; whose authority on any subject being taught cannot be questioned; and whose lessons are conducted at the right pace – a pace that is fast enough to keep every child interested and not too fast that some are left behind. The same teachers have relationships with the pupils that are good-humoured but never frivolous; purposeful but never strained.

So what are the secrets that these 'super-teachers' possess? Without a doubt, some of it is personality. A good proportion of it is in their planning and preparation. But a good proportion of it is also in the way they see and manifest their role as a teacher. This is what gives them class control that lesser beings would die for! This is what gives them the ability to form seemingly perfect relationships with every class they teach, regardless of how much grief these classes have given their previous, mere mortal, teachers.

On mixing business with pleasure

A young teacher visited a school before taking up his first teaching position there as the school's music teacher. The school was a tough Liverpool secondary boys' school where music lessons were not considered to be of any value by the pupils and an interest in music was considered to be 'soft'. These were the days before the National Curriculum and many music teachers believed that music lessons should concentrate exclusively on the study of 'high culture' and the singing of traditional folk songs – hardly the type of diet that would attract the average Liverpool or Everton fan!

The novice was asked by the outgoing post-holder, a middle-aged woman pianist and singer, who had a look of tiredness and defeat about her, whether he liked music. 'Like it? It always has and always will play a major part in my life,' he replied. 'Then you are in the wrong job,' she responded. 'You will hate it by the time you are my age!' This was the first indication to the newly qualified teacher that his chosen profession could be a dangerous place for those not prepared to be flexible in their approach to teaching.

What the 'old hand' was saying is that teaching music is tough – particularly if you have a narrow and conservative perception as to what 'music' is. Like the young mathematics teacher in the previous chapter, who saw inner-city children and not the mathematics as the real challenge, the correct course of action for the young music teacher is to educate his pupils in music from points and perspectives that are accessible to them. Thankfully, the National Curriculum now acknowledges this approach as being worth while and provides a clear and relevant framework for teaching music, which was lacking in the earlier days of teaching music.

Staying with music, consider the following strategy for appointing specialist instrumental teachers.

Teachers of children or teachers of instruments?

A manager of a Local Education Authority (LEA) music service was interviewing peripatetic instrumental teachers to positions in a music service. He asked them 'What do you teach?'

The candidates who had thought about their job and who were not simply filling time between professional engagements paused, knowing that there was a good chance that the manager would be able to recognize the instrument they were holding in their hands as (say) a clarinet. Most then replied, 'Children.'

Often these candidates got the job – even if they weren't the best instrumentalists being interviewed that day.

A good teacher teaches children, not mathematics or music, French or football. The ability to motivate, to capture children's imaginations, and to inspire them to learn for themselves are together the hallmarks of a good teacher. These individuals will never have discipline problems and they will always have the attention of their pupils. Quite often, in a secondary school, a child's favourite subject will be the one taught by their favourite teacher. It is all too common an experience for a child's love of a subject to wane when there is a change of teacher. A good teacher will have subject knowledge but will also have the ability to manipulate that knowledge so that the key points become interesting and accessible to all pupils. These teachers will plan their lessons and learning activities in such a way as to maximize pupil interest and enthusiasm.

The correct approach

Teachers make choices throughout each day in school. At the planning stage a choice needs to be made, for each activity, about which is the most effective way to organize learning in that activity. The purpose of this book is not to make judgements on the merits or otherwise of whole-class teaching, group teaching, paired work or individual work, but to emphasize that each strategy has a role to play in the effective teacher's repertoire of classroom organization. The decision to use one technique at a particular time will be based on the teacher's knowledge of how the class will react to tackling this particular task in that particular way. What is right for some pupils may not be right for others and so, for some pupils at least, changes in the styles of learning may help to keep interest and promote learning.

Teachers who are sensitive to the response of their pupils will also know when a particular strategy is not working as well as it should be. They will have a clear enough picture of what they are trying to achieve (articulated in the Learning Outcomes perhaps) to be able to switch tactics and try a different approach. If individual or small-group work in a lesson is not going well, the teacher may need to increase the number of 'whole-class interventions' in order to try to assist pupils in their understanding. If, alternatively, a whole-class teaching session is not going well, then it is likely that the concept being explored is too difficult for the pupils or that the teacher is not explaining it at a basic enough level. The presentation may thus require some simplification, or a change of focus of the explanation (hence the importance of good subject knowledge by the teacher) or a switch to an easier concept that will prepare the pupils for a higher level of understanding later on.

A good teacher will know when to make a change but such a teacher is also more likely to have considered the potential problems with a particular approach at the planning stage. A weaker teacher, in contrast, is likely to battle on regardless and, as the lesson progresses and pupils become more and more frustrated, runs a serious risk of losing control of the class.

The wisdom of Solomon

One other, equally important, attribute possessed by good teachers is their ability to be fair and even-handed with pupils at all times. All incidents in a classroom have a cause. Usually there is a child in the wrong and a child in the right; sometimes they are both in the right or both in the wrong. A teacher's response can be to punish or to educate. Children being children, unbridled use of rebuke or other forms of punishment means that there will be winners and losers being created in the room throughout the day. The winners usually delight in letting the losers know who is who, and this leads to tensions and undercurrents forming in the classroom.

Although there are times when a firm rebuke or other sanction in line with the school's documented practice is appropriate, a good teacher will make it clearly understood what the cause of the sanction is and why it is being implemented. The process is meant to educate other children as to what the acceptable boundaries of behaviour are, as much as to punish the one who stepped out of line. You will never see ridicule in the repertoire of control techniques used by a good teacher. Humour, yes – at an early stage perhaps; but ridicule belongs in the same category as oppression and aggression described elsewhere.

Children need the security of consistency in their teacher's response to misdemeanours in the classroom. Where there is consistency and fairness, there will be far fewer 'incidents' than where there is inconsistency and a pattern of different responses to different children on different occasions. In these instances, children learn that sometimes you get away with it, and so it is probably worth having a go. In well-managed classrooms, children know exactly what they can and cannot get away with and are more likely to comply with the teacher's own clear expectations and the class or school behaviour code.

This consistency, based on fairness helps reinforce good moral standards in children. Where they see adults bend the rules then they will learn that it is acceptable to bend the rules. Where they see adults comply with the rules then they will learn that this is the morally stronger approach to problems and difficulties faced in life.

Summary

A good teacher recognizes that the pupils, rather than the curriculum or subject being taught, should be the focus of effort. Good planning will help the teacher choose the right teaching and learning techniques for each task. A good teacher will keep the respect of pupils, and their interest in the lessons, through being fair and even-handed in keeping control. Children find security in consistency.

THE CLASSROOM

How important is the classroom to the success of learning? What does a classroom say about the teacher in charge? Will 19th-century classrooms serve the needs of the 21st-century child?

The classroom is the teacher's office and workshop, the premises where a teacher conducts his or her business. Any workplace has to be set up to maximize efficiency. The use of the available space needs careful consideration to ensure that it is being used in such a way as to encourage a high work rate, so as to make it easier to work.

Classroom or laboratory?

Classrooms are becoming increasingly complex places. There are still the tables or desks, and in some schools they are arranged in rows – not tradition for the sake of tradition but simply that many teachers of older primary school children and secondary school children appreciate that eye contact between teacher and class helps enormously in keeping the class engaged, while eye contact between a pupil and another member of the class invites distraction.

Classrooms still have a blackboard or whiteboard available, and in many primary classrooms there is increasing use of flipcharts. There is still, in most primary classrooms, the carpeted area for whole-class discussions or story reading. But most classes now have at least one computer (which should be in more or less constant use), and listening posts with tape machines and headphones, to aid readers. As more children with special needs are taught in mainstream education, other technology, such as CCTV or personal workstations, may also appear in some classes.

Good teaching usually occurs in classrooms where a visitor doesn't initially notice all this equipment. These are the classrooms where the people come first. They are rooms where the teacher can move around easily and has good sight lines and reasonable physical access to every child. They are the rooms where children can move around easily, if they need to, to get books or resources to aid them in their work. Often the layout of a room will contribute to the potential (or otherwise) for disruptive behaviour or for pupils losing concentration. If children know that the teacher cannot see them for extended periods of time, the temptation to become involved in illicit activities will eventually become too great to resist for many. Also, if children have difficulty going to look at a reference book because they cannot get out of their place without disturbing others, then often they will decide not to use these resources, with consequent implications for the quality of their work.

Does size matter?

Many primary school teachers complain that the size of their classrooms is inadequate. Often they are right, but there are many examples of good teaching occurring in overcrowded classrooms and of bad teaching occurring in the most spacious and generously proportioned rooms.

Take two contrasting school buildings, as described in the next case study, that were constructed within 10 years of each other but where the design was guided by very different perceptions as to how learning should be organized.

One school has pitifully small classrooms and, being a popular school, has quite large numbers of pupils in each class. The rooms are L-shaped and were never really designed for classes of around 30 children, each with their own table or desk space. At the time of building, the intention was probably that much of the teaching would be team teaching and would take place in the larger shared communal areas outside each pair of rooms.

The school is a two-form entry primary school and has developed very good practice in team teaching but still uses these rooms with sometimes more than 30 children in them for much of the school day. In these rooms *unnecessary* equipment, including bookcases, computers and even desks for the teachers, simply doesn't exist. Books are in the well-appointed libraries and the computers are pooled in the communal areas. Teachers have less space to accumulate stacks of unused papers and other bits and pieces, and so their 'personal' corners in these rooms are much more focused on what they really need. Sight lines can be a problem, but the teacher often operates from the corner of the L shape so that all areas of the classroom can be seen.

Informal discussions that would normally take place on a carpeted area are less frequent than elsewhere but can take place in one of the libraries. Because of the shortage of space, teachers have had to think out very carefully what it is they need in their rooms and what the key activities are that can take place in their rooms. The communal areas, used for practical work in science, design and technology and art, as well as for information technology, are carefully timetabled so that pupils from more than one class can sometimes be seen engaging in similar activities in these areas.

Another school had very generous accommodation, particularly for its infant classes. Thirty or more children can often appear lost in the large square classrooms. The infant rooms are divided into clearly marked zones, the most common layout being of two areas with tables for the pupils to sit at, a carpeted area of generous proportions, and a games/play area. This is in addition to a 'wet' area shared with other infant classes beyond a second set of doors.

Nevertheless, the rooms look untidy as piles of 'clutter' have been accumulated in corners. The rooms are big enough to accommodate the keeping of old work, and so old work is being kept in abundance. In one room a window display, of the 'stained glass' variety, was known to have remained unchanged for two years. Conversely, inexperienced teachers found that they had too much space and could not fill it meaningfully. The result was a permanently untidy environment that never looked 'finished'. Despite the large square floor plan, teachers often had inadequate sight lines because they used bookcases and storage units as dividers between areas.

The first school described above is a successful school that manages very well in difficult circumstances. Whether or not improved accommodation is likely to improve standards attained by children is difficult to assess because the teachers have developed ways of organizing their work that makes their current accommodation serve their needs. In fact, a change in accommodation could put at risk some of the organizational strategies that make this school successful – for example the high degree of teamwork that exists between teachers, and the high degree of focus on well-organized practical activities.

In the second school in the case study, regrettably the teaching often appears unfocused. All too often during lessons, small children on different activities appear to be a long way away from their teachers, and noise levels then rise significantly.

There are, doubtless, many instances where poor teaching is occurring in overcrowded classrooms and is, perhaps, in part being excused because of those physical circumstances. There are certainly also many instances where good teaching is occurring in well organized and generously proportioned large rooms. What makes a teaching area contribute successfully to a good educational experience is not its size but the way it is organized and used.

A teacher who sees the room as a place of business and who sets it up for efficiency of movement, visual and physical contact, and ease of access to relevant resources will make any classroom – small or large – work successfully. Combined with the type of attractive, relevant and meaningful display discussed earlier in this chapter, these rooms have the potential to be really effective places of learning for the children occupying them.

Looking to the future

The mushrooming influence of information and communications technology on our lives is inevitably going to have a significant impact on the classroom and the way we teach our children. Whilst it is doubtful, in the foreseeable future at least, that the computer will replace the teacher, it is certain that the relationship between the teacher and pupil will change as information technology is used more frequently in learning. It is not as much from the increased access to learning resources and interactive learning experiences that the biggest changes for teachers will come; the major difference will be in the use of information technology to:

- ❑ plan and organize work;
- ❑ set work for whole classes and for individuals;
- ❑ mark the work on line (even handwritten work can be scanned in and submitted for marking);
- ❑ record pupil assessments;
- ❑ share progress reports electronically with parents.

The Internet and, more relevantly, local Intranets will provide opportunities for home–school liaison and for inter-school liaison that could only be dreamed about previously. The size of classrooms in the final analysis may not matter, because children may only visit them for a short period each week, and even then in smaller groups. What *will be* important is for teachers and managers to accommodate today's technology properly in their classrooms and be receptive to the advantages of new initiatives in helping to resolve old problems in education.

Summary

Good organization of the classroom is essential to promote learning. Pupils need to be seen and to be able to see. Poor use of space can hinder learning and can contribute to loss of order in the classroom. The size of a classroom is not as important as the use to which teachers put it and the whole building.

Classrooms need to be 'technology-ready'. The future holds very real possibilities of significant change in the way we use our schools and teach our children.

CHAPTER SUMMARY

The heart of every successful school is the quality of its teaching. This chapter has highlighted the following points with the aim of preventing the influence of teachers tending towards failure amongst pupils:

- ❑ Good whole-school systems for planning the curriculum and for planning the day-to-day teaching can help teachers clarify their own thinking, identify what it is that they are teaching, and recognize when their teaching is at its most effective.
- ❑ Where these systems do not exist, teaching often has vague – if any – identified outcomes, and people should not be surprised in these circumstances if learning by children is hit-and-miss in nature.
- ❑ Children learn by example, and the quality of their teachers' own work, in terms of presentation as well as content, is important. When faced with poor letter formation by a teacher, for example, young children will have difficulty learning to write properly; and when constantly exposed to untidy board work or worksheets, spelling mistakes or poor grammar, children will never learn the importance and value of a pride in good presentation.
- ❑ High standards are not confined to academic work in good schools, and high expectations by teachers are not confined to neat and tidy work.
- ❑ Children like consistency and, within the bounds of appropriateness for age and activity, order.
- ❑ Good discipline is one hallmark of a good teacher and good class control is another. Both, however, are meaningless to successful schools if they are not built on respect, openness, tolerance and a morally sound value system operating within the school.
- ❑ The learning environment is essential to good learning and indicates that teachers should pay attention to organizing their classroom and their pupils to suit the learning taking place and in order to minimize disruption.

4

RECOGNIZING THE SYMPTOMS OF FAILURE (2): MANAGERS

Without a doubt, the single most significant factor influencing the overall quality of education in a school, and the single most significant factor in determining whether a school will succeed or fail, is the management. The management in this context usually means the headteacher, because he or she alone, through credibility of action and through the processes of involving others in management decisions, has almost complete control over the school's direction.

So how much power does a head really have? In a small primary school this is easy to see. The headteacher is the individual who, often single-handedly, runs the school on a day-to-day basis, leads the staff through meetings, takes many initiatives, is the focal point of contact for parents, and is the 'figure of authority' for children. In a secondary school a headteacher can have two or even three deputy headteachers and a team of senior teachers who, between them, will have a significant amount of non-teaching time to help run the school on a day-to-day basis. It is the management of this team, and how the work of this team interfaces and relates to the rest of the organizational structure of the secondary school that is important. The secondary headteacher can decide, like his or her primary counterpart, to delegate significantly or not to delegate at all; to share the decision-making processes or not to share them; to seek the views of others in moving the school forward or not to seek them; and to steer the ship with the help of others or alone.

Few schools will be deemed to be failing to provide their children with a satisfactory standard of education if they are well managed. Such is the nature of management in education that a school can only be well managed if the features that determine whether it is successful or not are being effectively addressed.

In this chapter the symptoms of failing management are examined and the attributes of good management are discussed, from taking responsibility for the state of the school through to the effective systems necessary to keep management in touch.

THE ESSENTIALLY HEADLESS SCHOOL

It is possible for schools to be appearing to succeed despite poor management. Sometimes the teachers in a small primary school will shut themselves off from an ineffective or negative management style and will get on with the job regardless. In the worst cases, the education being provided will have significant shortcomings as teachers continue their work without central guidance and without the benefits that being a member of a well-led team can produce. For example, teachers can be working in isolation without any strategic initiatives to review and improve either the curriculum or the organization of the school. The following case study illustrates a school running without a headteacher in anything but name.

> In one Midlands primary school, a struggling headteacher operated relatively ineffectively without a chance of ever really taking control of the school he had been appointed to manage. The school was already being efficiently managed by an unofficial committee of empowered young female teachers. Masses of energy and several strong personalities, together with an informally agreed 'middle management' structure, effectively excluded the newly appointed and hopelessly overawed male headteacher from establishing himself as an effective and credible leader and manager.

At first sight this arrangement might seem to work. The headteacher fulfilled many of the 'public' roles expected of him, taking assemblies and being available to parents who may wish to see him. But in practice he had very little control over strategic decision making, and the major resourcing and organisational decisions were taken by the 'middle management' group without reference to him – even though the legal responsibility for the outcomes of these decisions was his.

The middle management team was exercising power without responsibility whilst the headteacher held responsibility without power. The only course of action for this headteacher to take, if he wished to stay in post, was to begin to move his own position from one of figurehead to that of leader. However, he would have to do this without alienating his very able staff. He needed to bring the structure within which they worked into a whole-school management structure that provided a communication and decision-making system from top to bottom; and he needed to ensure that their energy and motivation was used to inform and influence strategic decision making rather than to decide it.

In a second example, the situation of running a school without a headteacher at all is illustrated – this time in relation to the governing body.

> A group of school governors, when faced with the loss of yet another headteacher in quick succession to the previous one and an increasingly critical parent body, colluded with three confident and ambitious female teachers to run the school, effectively by committee. The plan was to allow the three to share the role of headteacher. Legal problems and the swift intervention of a higher authority ensured that this idea did not develop and the school eventually appointed a competent head and deputy.

In this example the governors clearly felt that the leadership that the staff had failed to get from a chain of weak headteachers could be provided by the staff themselves. The account illustrates the frustrations felt by teachers who understood the need for a well-managed team in moving a school forward and who were desperate to work in such a context. The governors, having failed to appoint or develop a successful headteacher themselves, were at the point of effectively devolving this responsibility to their teaching staff by letting them run their own school!

Of course, such an arrangement – even if it were legally allowable (which it is not) – would be very difficult to sustain over time. Who would be ultimately responsible for day-to-day decision making or for decisions affecting pupils' health and safety that may need to be made at very short notice? What if the team disagreed? How would the school have handled personnel changes? Such an arrangement is fraught with difficulties.

THE HEADTEACHER IN DENIAL

'I see no problems. My school is fine.' This can sometimes be the cry of a headteacher who is facing serious difficulties in the school but who is in denial of the situation and does not accept that the problems exist. When a headteacher and the other members of the management team lose touch with what is going on in their school, the school's results will, sooner or later, become disastrous.

Present but not correct!

Some headteachers will survive in a school for years despite an ever-increasing body of evidence that they are simply not doing their job well. Often, these individuals have built up a rationale for explaining away every possible criticism that can be levelled at them or their school. These are the headteachers who steadfastly refuse to acknowledge that there is a problem even when it is staring them in the face. Surprisingly, they defend weak teaching. Some will make excuses for a poor teacher, saying for example that problems with health or in their personal lives are distracting their attention from teaching; they may assert that, because of this, their performance in their job is, temporarily, not up to its usual high standard. Others will question an inspector's evidence base, saying that it is not possible or fair to judge a teacher's performance on the basis of just one or two observations. They will, conveniently, not acknowledge the related evidence taken from an examination of such a teacher's planning, record keeping and pupils' past work. They challenge parents who dare to complain, unfairly using an assumed professional seniority to dissipate a parent's complaint. On other occasions, the same head may overreact to a complaint and cause unnecessary distress to a teacher or pupil through a mishandling of it.

These headteachers are often masters in disguising problems from the casual observer. To the occasional visitor, they can give the impression that a school is well managed and is successfully functioning without any difficulty. Try to dig a little deeper, though, and it is very difficult to uncover recognized management strategies or effective communication systems within their school. But begin to talk to staff and soon they will open up 'off the record'.

One feature of these schools is often a disenchanted deputy head, whom the staff continually complain to because they cannot get through to the head or because the head has no credibility in their eyes. The poor deputy, beleaguered and alone, often has no one else to ask for help. It takes a strong deputy to go to the governing body or to an external authority, because doing so could well wreck whatever partnership may exist with the headteacher. In a small school such moves, seen by the headteacher as disloyalty, can cause significant tensions and may well divide the teaching staff. Such is the nature of education that a deputy going down this route can never be quite sure what the final outcome of the 'whistle blowing' may be and may well finish as the loser!

There is little that can be done to help a headteacher who has developed these myopic tendencies – who refuses to acknowledge the existence of a problem. Not every school can depend on a band of willing staff to take over its management in these circumstances, and not every secondary school will have deputy headteachers who are willing and able to try and fill the vacuum. Indeed, it is often the case that a potentially good deputy headteacher will leave the school quite quickly when such a situation arises. This is often a tangible warning sign for governors and other authorities that there is a problem in the school, particularly if they have had concerns themselves but no hard evidence to date.

Accepting the problem

If problems in a school are going to be addressed, it is essential that those with the power to address them – and particularly the headteacher – recognize the existence of the problems. Without this recognition a good school will eventually decline and a satisfactory school will begin to show signs of failure. All the support that a local authority can provide and all the additional training that a governing body can pay for will mean nothing unless the existence of the problem is acknowledged. And without this acknowledgement there can be no real desire to change.

It is difficult to know what motivates individuals to deny the existence of a difficult situation that is glaringly apparent to other observers. Perhaps it is simply a reluctance to accept that, at the pinnacle of their career, they still have a lot of learning to do and that they are not the experts they may think themselves to be.

Some of the most rewarding experiences that those who support schools can have in education centre around schools identified as having weaknesses that have made well-managed and successful recoveries. When the headteacher has openly acknowledged these weaknesses and has begun, systematically and with the full co-operation of staff and governors, to address them, then the only outcome can be success. But it takes time and a lot of hard work to turn a school round. Years of neglect of school systems cannot be put right in one term; ingrained low staff morale cannot be raised overnight, for there may be a need for some personnel changes and this can take time; standards of attainment cannot be raised in a week – although it is amazing what can be achieved when a school really gears itself up, as the following example shows.

One school, faced with significant competition from neighbouring primary schools and an increasingly critical governing body because of low scores in national assessment tests, re-organized its teaching for a term and a half in the run-up to the next round of national tests. Year 2 and Year 6 classes were taught in smaller groups, organized by ability. This was achieved by using the teacher employed to support pupils with English as an additional language (a 'Section 11' teacher), the special needs teacher, the headteacher and the deputy headteacher to supplement the classes' own teachers. For most of the day during this period, the children concentrated solely on English, mathematics and science in their revised ability groups. The result was a spectacular increase in their national assessment test scores in all three subjects, which put the school among the better-performing schools in the local authority.

Without a recognition by the headteacher in this example that there was a problem and that the headteacher alone had the authority and responsibility for solving the problem, the spectacular results achieved could not have happened. More important than the high national assessment test scores achieved was the effect that the whole strategy had on staff morale. Teachers worked as a team, with specific focuses in mind, and the benefits worked to the good of the whole school.

It takes a brave individual in the headteacher's chair to acknowledge problems and then get on with addressing them, but for this headteacher and this school at least the rewards were great.

Summary

No problem can begin to be resolved unless people recognize that it exists in the first place. If a headteacher will not recognize the difficulties he or she is leading the school into, it becomes difficult for staff to operate effectively and there is then a risk of losing good teachers who would often rather leave the school than risk confrontation through 'whistle blowing'.

Conversely, when problems are recognized and tackled with strong leadership, the results can sometimes be surprising.

A FAILURE TO TAKE RESPONSIBILITY

Another type of professional denial employed by some weaker headteachers is to acknowledge that a problem exists but to excuse it and effectively disown it. 'We have always had that difficulty. That is outside my control,' they may say to their staff or outside observers.

At its worst these headteachers can persuade their governors that the problem is really unavoidable or that it is somebody else's responsibility. The problem then becomes corporately disowned and increasingly difficult to put right. It can be difficult enough for an authority such as an LEA or a diocesan authority to address a weak headteacher in a school with an ineffective governing body. It is much more difficult when the governing body is, for some reason, defending the headteacher despite the available evidence.

Weak management of supply teachers

One common example of this tendency to disown a difficulty concerns failing supply teachers. Supply teachers can be employed by a school for a number of reasons, as follows:

- They may be taken on to cover on a regular basis for teachers who are attending an ongoing course. The individuals employed in these circumstances are often known to the school and may even be offered some kind of part-time contract. Their regular presence gives the scope for good class teachers to agree with them exactly what is required to be covered, on a weekly basis, in the class teacher's absence.
- They may be employed to cover for unexpected short-term staff absence. This will usually be for one or two days only, and regular class teachers may or may not leave work for them. If the relationship with the class or school doesn't work out, there is little damage done and the supply teacher will probably not be used by the school again.
- They may be taken on to cover for long-term staff absence – for example, a planned maternity cover or an unexpected long-term illness. Sometimes supply teachers in these circumstances are offered a temporary contract for a half-term, a whole term, or more.

This third area is where problems can set in. Often, a career supply teacher can be a good appointment in these circumstances. Sometimes, though, schools rush in and appoint a supply teacher on a fixed-term contract without satisfying themselves as to the appropriateness of the individual to the job that needs doing. The fact that they have done well in a neighbouring school may be more to do with the circumstances they were working in and the job they were required to do rather than their ability automatically to succeed in this second school. More often, there is nothing wrong with the skills and experience that an individual supply teacher has to offer, but there can be a lot wrong with the way that the supply teacher is inducted into the school and then supported and managed on a day-to-day basis.

Supply teachers are as much members of the professional team as any other teacher in the school. As teachers, they will have received the same training as the school's regular class teachers. They will often be highly experienced, and are likely to be paid the same, if not more, than many other teachers. Yet there can be a reluctance by headteachers to manage them in the same way as other staff – a reluctance to treat them in the same way as the school's permanent teachers and to make them a part of the team.

Inspectors are often asked whether a weak supply teacher can damage a school's chances of a successful Ofsted inspection. It is as if the head-teachers asking these questions see the supply teachers' work as being separate from the work of the rest of the school – as if they see the children being taught by that supply teacher as being temporarily outside the school's real area of responsibility.

One particularly successful small school faced its Ofsted inspection with three supply staff. This school did nothing different in the way it handled these staff. It had a procedure for welcoming supply staff into the teaching body and saw no reason to change this procedure because of its forthcoming inspection.

The supply teachers were each given a full induction to the school's systems and common practices. This was reinforced by giving each of them a copy of the staff handbook, together with guidance notes for supply teachers and visiting teachers. These indicated who held key responsibilities in the school, where assistance could be found for particular problems, and other such useful information. The new teachers were included in staff rotas for break-time supervision and were expected to attend staff meetings, even being assigned to staff working groups. Most importantly they were expected to submit their planning for examination by the headteacher or deputy every week, along with every other class teacher.

Far from complaining that they were being 'taken advantage of', these teachers were being treated professionally and valued the welcome and quality of information that they had been given, together with the trust that the school invested in them. As temporary staff members of the school, they appreciated a full and appropriate involvement. After all, they were being paid on the same salary scale as the other, permanent, teachers and, being experienced teachers familiar with practice in a number of other schools, could possibly make a useful contribution to the school during their stay.

Sometimes a headteacher will say 'You can't expect a supply teacher to attend staff meetings, because it will put them off working here again,' or 'They are only supply teachers, and so you can't expect them to submit their planning for weekly examination.' Then the same headteacher will ask if the school can be held to account if one of these teachers is weak!

In a properly run organization, these questions would not even be considered. The supply teacher may be all that the school has available at a given moment to educate a class of children; looked at the other way round, the supply teacher may be all that a class of children has to educate them for a term or more. These are the children who, as much as the children in the class next door, deserve a high-quality education. Governors should not accept that a school may have a weak supply teacher, and parents certainly should not accept it. The management of staff is the sole responsibility of the headteacher, and it is the head who refuses to manage effectively and who makes excuses who is selling his children short and inviting trouble for his school.

Weak management of newly qualified teachers

Many schools adopt a recruitment strategy whereby they strive to maintain a balance between new teachers and more experienced staff. This means that a small primary school may take one newly qualified teacher every two years or so and that a large secondary school may take four or five – and sometimes more – each year. It is not uncommon for newly qualified teachers to experience some difficulties. What sometimes surprises outside observers is an attitude that some headteachers show when they agree that a newly qualified teacher is having difficulties: instead of addressing the problems, the head simply applies a broad-brush response bemoaning the quality of the work done by the training institutions and effectively disowns the problem.

As stated earlier in the chapter, under similar but distinct circumstances there is a need for a headteacher to recognise the problem, to accept responsibility for it and to devise a course of action to deal with it. New teachers represent, through their training, a significant investment of public funding. Schools are not managing this investment well if they expect new teachers to fit perfectly into their first positions without a proper support structure.

Difficulties experienced by new teachers can often be resolved if recognised and addressed at an early stage. Problems may be encountered where schools do not have a comprehensive induction process supported by regular monitoring and review of the planning and classroom performance of new teachers. The induction process is essential if new staff members are to know and use existing school systems effectively. The monitoring and review process is essential if best practice in every aspect of teaching is to be encouraged and promoted. Where schools have such structures in place, they are in a position to offer their new teachers every chance of overcoming these early difficulties before anything develops and becomes much more serious.

Weak management of the low-achieving pupil

More serious than failing to acknowledge weak teaching is a failure to take responsibility for low performance by pupils. All too often teachers and headteachers will effectively 'write off' the children they are responsible for teaching because of the geographical location they come from or because of their historical pattern of attainment. This is one of the most serious derelictions of duty that a teacher or headteacher can be guilty of and yet, surprisingly, is still evident in some schools throughout the country. The professional term for this is 'low expectations'. The actual damage done to children is to jeopardize their future chances of success, because somebody somewhere didn't think that these individuals could do any better.

The UK government's initiative on target setting will go a long way to address this problem, but only if the initiative is properly managed and sustained for a long enough period of time for it to be effective. In addition, the schools that are selling their children short will continue to resist rigorous targets, citing a range of associated factors as good reasons for doing so. Now, there is nothing wrong with a school wanting, when agreeing its targets, to take into account socio-economic factors, such as the number of pupils on free school meals or the number of pupils from homes where English is an additional language. However, close examination of schools with a similar profile – an essential part of the target-setting process – should ensure that schools adopt targets that are comparable to the best performance by similar schools. No other target can possibly be acceptable.

Remember the school in the case study earlier in this chapter that, through imaginative reorganization of its available staff, was able spectacularly to improve its national test scores in one year? When headteachers claim that a problem is beyond their control, what they are really saying is that they do not have the will, the ideas or the imagination to address the problem. For them, it is easier to deny that anything can be done than to take responsibility and resolve the issues.

Summary

Whatever conditions headteachers may face, those are the conditions that they have to manage. No problems within schools that affect pupils' learning are outside the headteachers' control. Accepting responsibility for the performance of all staff, including supply teachers and newly appointed or newly qualified teachers is the first step on the road to resolving the problems they may bring with them.

Children, too, deserve to have their own problems recognized – and deserve to be supported in overcoming them. They are not to blame for their apparent lack of ability or the area they were born in.

A FAILURE TO CHECK PLANNING

Even with a headteacher recognizing and taking responsibility for a problem, difficulties in a school will not be resolved unless the headteacher has, or can develop, systems to support improvement. Indeed, the absence of these systems is likely to lead to problems in a school in any case. One such important aspect is the monitoring of teachers' planning.

The importance of monitoring planning

We have already seen in Chapter 3 how systematic planning needs to be effective in ensuring that the work of a school is properly focused on the real needs of its children. Most schools have developed suitable systems for ensuring the long-term and medium-term planning of their work. This means that an outline is produced, say for a term or a half-term, showing what each teacher intends to cover from the school's schemes of work. We have also seen how short-term planning, on a day-by-day or week-by-week basis, is important for identifying measurable learning outcomes for different groups of children. Most schools will recognize this practice as good practice and will, after some examination of available formats, introduce a whole-school format for short-term planning.

However, choosing and introducing a planning model is really only half the story. In order for the model to be effective, it needs to be used consistently throughout the school year and modified, if necessary, to suit the specific requirements of the school in planning. Above all, it needs to be used in the same way by all staff.

In some schools, once the short-term planning model is introduced, that is the end of the initiative as far as the management team is concerned –

and often the end of the initiative as far as some of the teachers are concerned too. Where headteachers leave the short-term planning solely to the individual teacher and make no attempt to ensure a consistent practice in planning, the benefits of the initiative are likely to be lost. Some schools have worked hard to introduce planning models, only to have them shelved by individual teachers who have found 'something better', or customized by other uncooperative teachers to the point where the models are not really worth using.

Quite often, the reasons cited in primary schools for not monitoring planning include lack of time by the headteacher or, in some cases, no desire to question the professionalism of staff by continually looking over their shoulder. There are strategies that can be used to minimize the amount of time spent on monitoring planning if that is the issue, but lack of trust shouldn't come into it: the monitoring of planning is not about trust. Of course, it is one way to check that teachers are doing what they have agreed they will do but, more importantly, it is a way of management finding out first-hand that a relevant curriculum for all children, regardless of ability, is being continually planned and that pupils' previous attainments are being taken into account in lesson planning.

In one primary school where time was particularly short for the headteacher a checklist was devised that could be inserted into the front of each teacher's planner (see Figure 4.1). The list included sections for long-term and medium-term planning and a section for weekly planning. Each week, the headteacher would tick in a box in the appropriate column to show that the items on the list had been satisfactorily covered.

The method employed shows teachers exactly what is being monitored, shows them that the head has actually looked at their planning, and shows the headteacher, over time, what aspects of planning may be problematical for groups of staff or for individual teachers. It is possible for a headteacher who uses such a monitoring form, having a systematic and focused way of examining teachers' planning, to complete the exercise of monitoring planning in around half an hour each week. Similar strategies could be used in larger schools where some of the monitoring of teachers' planning is delegated to others.

Without such an initiative, the planning of teachers' work in the short-term can be very much hit-and-miss. A headteacher will be aware that there are problems in some classrooms, and some of this awareness will be gained from comments from parents who are unhappy with some of the work their children are being expected to do. In other instances, some teachers will express, in confidence, reservations about some of the occurrences in the class next door. The headteacher will have to look at the work of pupils and may have misgivings about what is being taught in some

Planning Monitoring Sheet		**Teacher** ______ **Class** ______									
Length	*Aspect*	*Notes (weak subjects, particular strengths, etc)*									
Long-term	NC coverage										
	Link to scheme of work										
	Link to previous work										
Date long-term planning checked ______ **Signed** ______											
Medium-term	Link to long-term										
	NC coverage										
	Balanced subject content										
	Broad Learning Outcomes										
	Assessment strategies										
	Differentiation										
	Major resourcing (inc visits)										
	Key organizational										
Dates checked 1. ______ **By** ______		**2.** ______ **By** ______									
Short-term (tick as appropriate)	**Week**	**1**	**2**	**3**	**4**	**5**	**6**	**7**	**8**	**9**	**10**
	English										
	Mathematics										
	Science										
	ICT										
	History										
	Geography										
	Design Technology										
	Art										
	Music										
	PE										
	RE										
	Link to medium-term										
	Learning Outcomes										
	Differentiation										
	Assessment activities										
	Teaching strategies										
	Class organization										
	Use of support										
	Resources										
	Visits										
	Homework										
Notes											

Figure 4.1 *A sample Planning Monitoring Sheet*

classes but be uncomfortable about homing in on just one or two teachers. In contrast, once a new short-term planning model has been agreed and introduced, it should be a relatively easy task to undertake the monitoring of short-term planning as part and parcel of a new planning initiative.

Before a headteacher begins to monitor progress using a new planning model, a uniformity of approach by the teachers involved needs to be agreed. Otherwise, whilst very conscientious staff members might write out in full detail every aspect of the lessons they intend to teach and might create new categories on the sheet for additional information that they thought necessary, in contrast others might leave whole sections of the sheet blank and might put the barest detail in other sections – quite often the missing detail being related to how the teacher intended to address the different ability levels within a class.

The monitoring of planning doesn't stop any teacher from preparing in the amount of detail that others may find unnecessary. If that is what the teacher needs to do a good job, then so be it. Such detail can also be used as one of a range of exemplars for colleagues who are struggling with their planning. What the monitoring does do is ensure that every teacher is planning to a minimum standard. In addition to informing management that the schemes of work are being delivered, the planning can also be used on the occasions when a new teacher or a supply teacher has to take the class: it will inform them exactly of the nature of work being addressed with each ability group over time.

It is difficult to imagine that in some schools nobody ever looks at the short-term planning of the teachers. In such circumstances, none of the management team can be certain that what is being taught is relevant and has been properly prepared. Above all, nobody can be certain that the differing needs of children in individual classrooms are being properly addressed.

The monitoring of planning alone, however, does not tell the whole story and can only tell the headteacher or other manager what it is *intended* that each teacher will teach. Monitoring of teaching and learning itself is an essential strategy in ensuring a consistency of quality in the educational provision throughout the school.

Summary

Good teaching hinges on high-quality planning. How can a headteacher be confident that the best interests of every child in her school are being served if she does not monitor planning? Where a school adopts the good practice of having an agreed system for planning that is used consistently by all staff then monitoring can be easy. Monitoring allows a headteacher a 'way in' where she feels there may be problems and can therefore be a route by which quick support can be given to teachers experiencing difficulties.

A FAILURE TO MONITOR TEACHING

A second area where there are often weaknesses in a poorly managed school concerns the monitoring of teaching and learning. This involves direct observations in the classroom, and it is perhaps unsurprising that a headteacher who lacks confidence in his or her own abilities will be reluctant to try to measure the abilities of others.

'The captain's on the bridge'

We have all come across, at some point in our dealings with schools, headteachers who appear totally office-bound. Many of us are familiar with headteachers who are so involved in day-to-day administrative tasks that they forget that the main business of the school is taking place in the classroom and not in their office. Once again, these individuals would cite shortage of time as being the main reason for them never visiting classrooms.

However, the reason is not really shortage of time but the headteacher's willingness and ability to prioritize tasks and use time efficiently to address these priorities. Above all, it is a comment on how they perceive the really important elements of their job. The headteachers who understand the value of regularly scheduled visits to all their classrooms understand what the real priorities are in a school. These are the headteachers who acknowledge that the most important transaction taking place during the school day is taking place between the teacher and the pupil in the classroom.

'Why is everybody looking at me looking at them?'

Most managers feel an initial sense of discomfort in sitting in on a colleague's lesson and making notes. The discomfort arises for several reasons. Firstly, it may arise because of not really having a rationale for being there: common thoughts are 'What do I write?' or 'Where do I start?' or 'Do I really need this black, bound notepad? I feel so conspicuous with it.' Secondly, discomfort may come from feeling uneasy about sitting in judgement on colleagues: 'Could I have really done any better myself?' and 'Are my observations simply nit-picking, or are they worthwhile comments that will help the teacher improve in practice?' are common expressions of this feeling. Rather like the monitoring of planning, the monitoring of teaching needs clearly thinking out if it is going to be a useful exercise.

In one school encountered, the headteacher knew that standards attained by pupils were too low. He knew that performance of his teachers varied from class to class. Even though they were all dedicated and caring individuals, some were simply more effective than others in the business of teaching children. The head agreed with his local education authority that one way of improving practice was to undertake an ongoing cycle of lesson observations: initially each teacher would be seen every half-term (one a week in this seven-class school); and, over time, early repeat visits would be made to classrooms where there were concerns, whilst other teachers would be seen just once a term.

The headteacher found the criteria used by Ofsted very helpful in devising a checklist, but he also undertook training, in the form of paired observations and paired feedback, from an experienced LEA inspector. This was essential in ensuring that he was fully confident in recording his observations and consistent in the quality of feedback he was giving to his staff. The strategy was explained to teachers and the initial cycle published so that all could see that it was a fair system, with senior teachers being the first 'under the microscope'. Repeat visits were made by private arrangement with the teachers concerned and were not publicized to the rest of the staff.

This headteacher found that the benefits of this systematic monitoring of teaching was very valuable. Teachers all received detailed feedback, and notes were kept both of the observation and the ensuing discussion. In any follow-up discussion, the headteacher and individual teachers agreed targets, or areas for development, and these were revisited at the time of the next observation. Where there were real weaknesses, additional support was offered, in the classroom or, alternatively, in the form of additional training or the opportunity to see other teachers at work. In addition good teaching was acknowledged – so many times the 'negative' or critical side of any initiative that measures performance in education is the only one that is recognized and the many benefits that can come from these systems are ignored.

Staff at the school came to trust the system and see it as a part of the overall strategy for moving their school forward and raising standards achieved by children. One or two teachers were encouraged to rethink their careers or their position as a consequence of the increased monitoring, professional discussions and target setting, and it was clear that the school was a better place for them having done this, with the children in their old classes receiving a far better standard of education than they would have done had the situation not changed. Furthermore, other teachers improved less satisfactory aspects of their early practice and became recognized as effective individuals with bright futures in the classroom.

One of the features of this example is the nature of the relationship between the headteacher and the LEA in recognizing and resolving the problem. The LEA had raised concerns about standards and progress in a supportive but unambiguous way. The headteacher had recognized these concerns as valid and received training from an LEA inspector to help develop his monitoring, skills and strategies for follow-up. The LEA had fulfilled a responsibility to identify and challenge weak performance but had also fulfilled its role in effectively supporting the headteacher to help resolve the problems. The headteacher had accepted that the responsibility for resolving the problem though lay firmly within the school.

The system that is used to monitor teaching and learning at the school described in the case study is similar to the one in Figure 4.2. More detailed criteria for monitoring the quality of teaching and learning can be found in the Ofsted Handbook (published by The Stationery Office). Handbooks are available for primary, secondary and special education.

Avoiding unpleasant surprises

The risks of not regularly monitoring the quality of teaching are large. Where schools have been identified as having serious weaknesses, or as failing to provide their children with a satisfactory standard of education, it is often the case that the management team of the school has no way of informing itself as to what is going on in the classroom. This situation relates to the managers who often appear most surprised when they learn that there are weak teachers in the school. They are the

Classroom Monitoring Checklist	
Date of Visit	*Class*
Aspect	Notes
Planning:	
– clear Learning Outcomes	
– clear structure	
– differentiation/extension	
– organization	
Teaching:	
– reflects planning	
– well-paced and timed	
– appropriate/varied strategies	
– well-timed interventions	
– good support	
– accessible spoken language	
– appropriate resources	
– accessible written language	
– clear worksheets	
– good board work	
– set homework	
Learning:	
– pupils on task	
– understand instructions	
– use resources well	
– use dictionaries	
– use reference material	
– seek assistance	
– work well in groups	
– work well independently	
– work at good pace	
– behave well	
– complete homework	

Figure 4.2 *A sample Classroom Monitoring Checklist*

ones who often feel let down by colleagues whom they were trusting to do a professional job.

Such individual managers sometimes feel a great sense of injustice that their own reputation is being placed on the line because of the performance of some weak teachers in the classroom. Sadly, they have missed the point. As headteachers they are ultimately responsible for the quality of the work being carried out by every teacher in their school. With that responsibility, they have the authority to do something about it when a teacher is under-performing. In order to take action, they must develop and maintain the systems for actually identifying good and bad performance in the first place.

Summary

A school is not a ship that can be sailed from the bridge alone. A head-teacher needs to know what is going on in each classroom. The only way to know this is by going and looking. How else can a head tell whether the planning records examined every week really translate into practice?

The monitoring of lessons is an acquired skill and training helps a lot. Monitoring must be done to a system and must be followed up by feedback to the teachers concerned. If a headteacher does not know whether or not a teacher in the school is struggling, the headteacher must take personal responsibility for that lack of knowledge.

A LACK OF MANAGEMENT SYSTEMS

A third area that may need attention is the maintenance – or in some cases the establishment – of management systems in a school. These are the interfaces between head and deputy, between management and staff, and amongst other parties. Without these interfaces there is a lot that can go wrong and the head may never know.

Everything has its place

The best-run schools operate like well-oiled machines. There are routines in these schools that ensure everything that should happen does happen and also that ensure everybody knows everything they need to know in order for them to do their job effectively. However, even when schools appear, on the surface at least, to use the same routines and systems, why do some work so much more efficiently than others?

Let us start by looking at a fairly typical repertoire of systems that a school may have in place:

- ❑ First of all, there is the senior management team. In a primary school this may be as small as the headteacher and deputy headteacher, or it may include several other 'senior teachers' such as those responsible for key stage teams. In a secondary school the senior management team could include deputy headteachers and senior teachers, as well as other staff with key responsibilities.

- ❑ There may well then be a 'middle management' layer if the size of the school warrants this. In some primary schools there could be 'Phase Groups' or 'Faculty Groups', where teachers with a collective responsibility for a cluster of subjects or aspects have the opportunity to meet. Secondary schools may have more formal 'Faculty Groups', in which those individually responsible for subjects may be grouped and in which some larger subject departments may meet separately.
- ❑ There may also be pastoral groupings, phase groupings, subject groupings, theme groupings or other rationales for bringing staff together.

In a good school, the one thing that these groupings will have is accountability. Each of these groups will be responsible for identified aspects of the school's performance, and there will be ways of reporting back to management – and back to the rest of the school – progress on projects or areas that they are responsible for. Some groups will have been set up specifically to manage certain new initiatives; for example, in a primary school they may be responsible for evaluating the school's provision in history and geography and revising the school's scheme of work and resource base for these areas; and in a secondary school they may be responsible for reviewing aspects of pupil behaviour and evaluating the school's rewards and sanctions procedures. Other groups may have ongoing responsibilities, for example for assessment or for literacy. Each of these groups will be responsible for fulfilling certain tasks by certain dates, each of which will usually be identified in the school development plan. The groups will have reports to present to staff meetings or management meetings, and often to governors' meetings (or sub-committees thereof), and these reports should provide accountability and demonstrate how the groups are fulfilling their designated functions.

This type of practice should ensure that the developmental work of the school, and the ongoing maintenance of the curriculum and essential school systems, is fairly shared out amongst staff. It should also ensure that the priorities of the school development plan can be met through teamwork and that as many staff as possible have some ownership in the running and development of the school.

No place for anything

Where systems are less well established, the headteacher often appears overworked, stressed and uncertain. Groups may exist in name and indeed may meet regularly, but nobody will have really clarified their terms of reference or the criteria to which they should operate. There is no way of recording business conducted and no way of reporting back to colleagues on work that is being undertaken.

> In one school, inspectors found that there were wholly inadequate systems for managing initiatives. The school, despite the initial absence of evidence, insisted that records of all meetings were kept. The inspection team visiting the school insisted on seeing them if they existed. Eventually the team was presented with a large bundle of rolled up flipcharts containing notes taken at different types of meeting. The charts were normally stored behind the armchairs in the staff-room and, after retrieval, were examined and labelled by the school staff during the actual inspection for presentation to the inspection team.

The sad thing about the example is that the school had indeed been doing the work it claimed to have been doing but, unfortunately, no one thought it valuable enough to write it up properly and circulate it to staff either at or before the next meeting. There were good ideas in the flipchart notes – but ideas that had subsequently been forgotten or simply not acted upon, and ideas that could have shown the school in a much better light had they been developed. The creative energies of the staff in developing initiatives were not being properly harnessed or used.

The outcome of a school development exercise in this school was merely the initial discussion itself and there was no effective follow-up. Now, this is not to advocate note-taking for the sake of note-taking; but it is important when a professional discussion takes place that somebody synthesizes the discussion and ensures that the rest of the attendees at the meeting have the opportunity to reflect on the points made and respond to them. This is how ideas are developed and how effective action can grow from simple comments in meetings.

If a school can decline to the point where the headteacher has no record of what is being discussed at the various meetings that may take place or, even worse, if the school never had these structures in the first place, then the outcome is likely to be serious. Complex organizations like schools will only function successfully if everybody's energies are being used to pull the school in the same direction. There needs to be a central point of focus – a central control mechanism – that understands what tasks each of the various groups in the school is engaged in and how relevant those tasks are in helping the school to achieve its developmental aims. There needs to be a way of increasing the work of any one group, because its success may be imperative to a particularly pressing need that the school has, and scaling down the work of another if that work is seen as having a lower priority. In a small primary school, the headteacher will fulfil that function and will control and balance the work of different teams to ensure that the school's priorities are met. In a larger secondary school, the senior management team will fulfil that function but the headteacher will still have overall responsibility for ensuring that the systems in place for fulfilling the school's priorities are the most appropriate systems.

Summary

Good internal systems for decision making, developing initiatives and consultation are essential if a school is to harness the creative powers of all staff and give everybody a stake in the school's successful development. The absence of such systems can mean a lot of lost opportunity in a school that is, at best, likely to stagnate.

AN INABILITY TO COMMUNICATE CLEARLY

A fourth area where weak managers fall down is in the area of communication. This is more than having the formal systems described in the last section; it is about using them effectively in conjunction with other, sometimes informal, systems so as to ensure that everybody is as informed as they need to be to work at their most effective.

The basics of communication

Communication is the key to success in any organization, team or partnership, and in the final analysis education is about communication: communication between teacher and pupil is the element that triggers learning.

Communication is playing an increasingly important role in the lives of many young people as the use of computers in the home and in the school becomes commonplace. When even the youngest teachers were still at school it was unheard of for them to 'log on' to a computer and hold a real-time conversation with a school child from (say) Kentucky. Yet that is the type of world that children are growing up in – where the ability to communicate is also as valuable and powerful as the knowledge shared through that communication.

Of course, interpersonal communication in the 'physical' world requires different skills from electronic communication over the Internet, and it can be argued – justifiably – that the ability to communicate personally must be taught and practised before electronic communication skills are developed. In schools, much of the time spent teaching, when not on the subjects of the curriculum, is spent developing personal and social skills; and many of those skills are about communication. They are about listening to others and hearing what they say, about trying to understand how other people perceive a viewpoint or a situation. They are about talking to others and responding to their feelings or opinions. Examination results count for a lot, but the deciding factor between two equally qualified young people in a job interview will be their relative ability to communicate. This is a pattern that starts with the first interviews for holiday work or for college places and continues throughout an individual's working life.

So, if communication is such an important factor in education and in our wider lives, it is reasonable to expect that the best examples of effective interaction between individuals and within groups of people can be found in our schools. After all, schools are about people, and they are about the interaction that takes place between different groups of people. There is no other significant process involved, other than the process of individual learning.

Formal and informal communication in schools

Communication in the management of schools takes a number of forms, as occurs in any organization. Written communication is used to inform staff of long-standing arrangements or procedures. Formal meetings are used to present new initiatives, or to update staff on short-term developments. Smaller group meetings are used to deal with the business that specifically involves those groups. Individual meetings may be used for more personal interaction, such as appraisals or to conduct specific items of business.

In addition, there is an informal mode of communication, often individual and rarely in groups any larger than two or three, where ideas or opinions can be tested or circulated prior to formal introduction. These informal occurrences are often vital for ensuring the smooth running of a school and for developing the confidence and trust of staff. Informal communication allows ideas to be tested and developed, without commitment and without the risk of formal rejection at a meeting.

However, there must come a time when a proposal, tested informally in confidence perhaps with one or two colleagues, will be put to the whole staff (or other appropriate group) for full debate and consideration. It is important that this is not seen as a *fait accompli.* If it is, it is because too much debate has taken place informally before the proposal was presented to staff. Good managers will know who it is appropriate to discuss an idea or proposal with and how long those discussions should be kept going before taking the idea or proposal to the whole staff.

Poor-quality communication

Sometimes a headteacher does not have the quality of relationship with his staff to test ideas informally before 'going public'. These are usually the headteachers who find communication difficult. Of course they can write memos – and probably do in abundance! They can address a staff meeting, perhaps, and give clear instructions. But they have real difficulty in eliciting responses from people; difficulty in hearing what people are saying and understanding what they mean; and difficulty in seeing the responses of others as constructive and useful contributions that can help move an idea forward.

The problems that occur in these circumstances can be troublesome for a school and can seriously impede school development. Because the headteacher will only communicate at meetings, there is a high risk of failure when attempting to introduce new initiatives. Headteachers who lack interpersonal skills do not usually carry the respect of their teachers. New initiatives are more likely to be either rejected or only half-heartedly adopted in these circumstances. A cycle begins to create itself when, because of the risk of failure, the headteacher becomes reluctant to present initiatives. The school gradually begins to slow down as change becomes more and more difficult to initiate. It becomes difficult for others to develop new ideas as the culture of the school becomes one of caution and criticism. Often teachers will shrink back into their classrooms, share less, and develop their own practice in relative isolation.

The same headteachers find it difficult to deal with problems amongst their staff. Poor or abrasive interpersonal skills on the part of the headteacher can often be used by a weak teacher as an effective defence that puts the root cause of the difficulties as a 'personality conflict'. This often makes it more difficult to take action against a weak teacher, because the headteacher's own mistakes in dealing with the problem may well have prejudiced any possible outcome.

Sometimes – and often in primary schools – headteachers gain office through highly effective classroom practice and with little evidence of, or experience in, people management. In many cases skills can be acquired, and there are some useful training initiatives available for new headteachers; however, management training takes time to permeate through into natural and effective management practice. Initiatives such as the National Qualification for Head Teachers, aimed at aspiring school managers, should, once refined, offer some help in preparing individuals for headship.

Nevertheless, there are individuals in post – and there will doubtless be others in the future – who, despite training, have difficulty in using their acquired management knowledge to influence their management practice. Poor communication skills are usually at the heart of the problems that these individuals face. Sometimes, a manager will try to hide poor communication skills behind delegation, for example by asking a deputy head to take staff meetings. This strategy can never be entirely successful as the staff need to see their leader leading discussions and making decisions based on consultation.

Communication between headteacher and governors

One of the most important lines of communication in a school is between the headteacher and the governing body. It is the governors, after all, who are ultimately responsible for the school's performance. In a successful school there will be a clear understanding between governors and their headteacher as to the quality and timing of information required. Governors will generally have developed the practice of in-depth examination of information presented to them and will not be afraid to ask for further clarification or to challenge things that they are unhappy about.

In some schools, however, governors have difficulty getting information, let alone accurate information, out of their headteachers. Sometimes, the reluctance to challenge a qualified and highly paid professional causes governors not to press for information they need. Often, in these circumstances, the only information a governing body receives is the information that the headteacher wants it to receive and the only interpretation of this information is the specific interpretation that the head wants the governors to see.

Where, for example, disappointing national assessment test results are openly shared and acknowledged as being disappointing, there is a much better chance of an open and constructive relationship with governors, which could assist in resourcing an imaginative response to the problem, than where the test results are explained away because of a 'glitch in the cohort' or some technical reason.

The government's initiative on target setting requires all governing bodies to set targets for end of Key Stage 2 attainment in primary schools and end of Key Stage 4 attainment in secondary schools. Targets to be achieved two years ahead will be published annually and governors will also be required to publish their school's performance against those targets. In addition, the government has set national targets for literacy and numeracy to be achieved by 2002. This process of target setting will ensure that governors are given more information and that there is an obligation on them to discuss various interpretations of that information. For most schools this will not be a problem, but some governors may need support and encouragement in analysing and challenging their headteacher's particular gloss on a set of statistics.

Summary

Teaching is about communication. Schools should be the centres of excellence for successful communication. For a headteacher, interpersonal skills are essential. Formal communication has its essential place but so does informal communication. Informal and formal communication will keep both management and teachers informed and involved.

Governors must ensure that there are well-established lines of communication on all key aspects of the school between the head-teacher and themselves.

A FAILURE TO TAKE ACTION: THE 'SOCIAL FACTOR' IN SMALL SCHOOLS

Finally, in this chapter, it is necessary to emphasize that a school can have all the systems for monitoring planning, teaching and learning in place, and the headteacher may know the strengths and weaknesses of all his teachers, but all this counts for little if the head is not prepared to bite the bullet and take action on all problems uncovered. Headteachers are particularly vulnerable in this area in small schools where there is a strong 'social factor' amongst staff.

Small schools and difficult decisions

One of the most difficult obstacles for the headteacher of a *small* school to overcome is that presented by the 'social factor', where a closeness of social relationship develops amongst a group of people who are working closely together. The social factor may be a product of a largely successful team that gels well and carries over its relationships to social events; or it may be a core group of friends within a larger team of teachers. Either way, in a small school, there can exist a delicate social balance. This balance is often seen, by those on the inside, as being essential in maintaining a good working relationship, where any threat to this balance could upset staff and could damage these relationships. The perceived outcome of this situation is that the school is too small and too fragile to withstand the trauma of disharmony amongst the staff.

The fact that such a school has got to this position in its thinking is a concern in itself. It means that, at some stage in its development, the authority of the headteacher has been put at risk or compromised because social factors have been deemed to be more important than professional factors.

> A manager of a small team of peripatetic teachers had to ask one staff member to undertake a specific task each week in preparing equipment for an activity he was helping to run. The task was a little onerous and others helped when they could but, nevertheless, it needed doing and this individual was the one who could do it most easily. Sharp words ensued when, after the request for this task to be undertaken was turned down, the request was communicated as an instruction. The response 'Well, I'll have to choose my drinking partners more carefully in future then!' was made.

The peripatetic team, because of its size, did have a large social element. Much of the work carried out by the team was done in the evening, and it was not uncommon for staff to go for a drink together afterwards. The staff member was trying to use the social lever as a way of favourably influencing his working conditions. The manager's perspective was simple and correct: the job came first and what happened in the pub afterwards as a result of that was not a part of the job and had to be dealt with separately.

Giving way to this kind of pressure can be dangerous, as the following example illustrates.

> There was a headteacher of a small primary school who could not bring herself to criticize her teachers. Even when faced with overwhelming evidence of a weak teacher and a rising number of complaints from

> parents, the headteacher would say 'It is so difficult in a small school to single someone out in front of her colleagues and friends', or 'We are all friends here and the friendship is what makes the school successful'.

Unfortunately, the school was *not* successful and it was the friendship that was causing the problems. Eventually this headteacher began to ask for help from LEA advisers. It soon became clear, though, that the advice they gave to her was not being acted on. What she really wanted was the LEA advisers to do what she perceived as 'the dirty work' for her. Criticism from an LEA adviser could be seen as something that invites sympathy from colleagues and, whilst still needing to be addressed before the adviser visited again, would not necessarily jeopardize the headteacher's own relationship with the teacher.

This strategy wasn't particularly successful, because the headteacher found herself giving more than sympathy to the criticized colleague and this effectively undermined the work of the advisers she had called in to help. The result was that the difficulties remained unresolved and the teacher under review believed that she had the support of her headteacher and that the criticisms made by the LEA adviser were largely unfair. And this arose all because the headteacher placed what she saw as being a delicate social balance in the school above her professional responsibility to manage the staff properly for the best interests of the children.

In situations like this, problems invariably become worse and the social fabric that was seen as important becomes worn and weak as teachers recognize that what they are witnessing is poor, not good, management.

Small schools and decisive action

Recognizing that the main aim of a school is to educate children and not to provide a comfortable working environment for all staff at all costs is not only possible in a small school but essential. Take the following example from another small primary school.

> The headteacher of a small primary school was seen to maintain good relationships with all her staff but also maintain a professional distance that allowed her to act decisively and to talk frankly and critically if the need arose. Like many schools, there is

> a social programme in which the headteacher plays a part. The social programme, though, is not seen by the head as an end in itself. Whilst recognizing that it is 'down time' for the staff, it is also seen by the headteacher as a staff-development or team-building exercise. Of course it isn't labelled like that and there is nothing formal about it beyond the social conventions being indulged in at the time. Staff have a good time but the head is careful to make sure that she keeps professional issues well away from these occasions.
>
> The school functions very successfully. Staff understand why decisions are sometimes made that are not popular with all of them. Everybody knows that the hierarchy that counts in the running of the school is the hierarchy through which staff members carry responsibilities. Staff here would not dream of trying to use social levers to influence professional decisions that they may not personally find agreeable.

Summary

Some headteachers will have all their monitoring systems in place but will go to great lengths to avoid actually tackling a difficult situation with a staff member head on. This is often so the case in small schools, where the social relationships developed can get in the way of seeing through professional responsibilities.

It is important for headteachers to be clear in their own mind where the real priorities of the role lie and where professional and social boundaries should be drawn.

CHAPTER SUMMARY

There are a number of management issues that can arise in schools and that the management team will or will not take responsibility for. These are summarized as follows:

- ❑ Accepting that there is a problem is the first step to finding a solution.
- ❑ A feature of many unsuccessful schools is that the headteacher does not see the issues or does not accept that they are significant when they are pointed out.

- Governors are still, despite league tables, largely reliant on their headteacher being open with them. Even giving governors more access to data will only work if they are given an honest and realistic interpretation of that data by the management of the school.
- If the head says that perceived problems are 'not significant' or are 'being dealt with', or states that the problems do not exist at all, then governors in potentially weak schools will often accept such statements rather than challenge them.
- Sometimes the headteacher, teaching staff and governors will state that a particular problem is a distinctive, unchangeable, feature of that school or is out of their control.
- A refusal to take responsibility is the first sign of a management's inability to find strategies and solutions to resolve problems.
- The absence of essential management systems for monitoring the quality of preparation and planning, the quality of teaching, and the attainment of pupils can contribute to gradual declines in expectations, professionalism and, ultimately, pupil attainment.
- Schools are dynamic organizations, whose success is dependent on good professional relationships, common understanding and goals, and teamwork – features not found in weak schools, where communication and consultation systems can be either inadequate, inappropriate or non-existent.
- One particularly common problem leading to the failure of small schools is rooted in a manager's reluctance to disturb the social harmony amongst the staff by challenging weaknesses amongst colleagues who work closely together each day. This inability to separate the professional obligation from the personal relationship is potentially very damaging.

5

RECOGNIZING THE SYMPTOMS OF FAILURE (3): GOVERNORS

This chapter examines the role of governors in a school. Once the importance of the role is established, the importance is considered of sharing information and of governors coming to their conclusions about issues in the school. Good financial awareness by governors needs matching awareness and understanding of the factors influencing attainment.

THE NEED TO BE INFORMED

People become governors for a variety of reasons but, once they have taken on the role, it is fair to ask whether they are really ready for the responsibilities and commitment that the position involves. Being a governor means ultimately being responsible for a school. To be responsible for something requires having good knowledge about its condition, strengths, weaknesses and the way it should develop.

Why be a governor?

What is it that makes individuals want to become school governors? When this question was asked of several experienced and long-standing governors, each had his or her own reason. For some it was personal, for others political, and for a few professional. All had a desire to serve the local community through their actions.

There are a number of different categories of school governor, as follows:

- *Parent-governors*, who are often those within the parent body who could influence others and who have proved themselves through, for example, organizational work on the parent–teacher's association.
- *Local authority governors*, who are often elected councillors or political nominees active in their own political parties. Their local education authority – and whether these individuals are majority or minority representatives, it is still their LEA that is relevant – is ultimately responsible for the success of this school – through its funding, for monitoring and supporting the work of the school, for many of its key policies.

- *Teaching and non-teaching (staff) governors*, who are elected by their colleagues and provide a valuable professional input into the discussions of the governing body. It is the teacher-governors who, with their colleagues, have to implement the policies of the governing body and who are in the front line, with the headteacher, on occasions when governors ask probing questions about the school's performance.
- *Co-opted governors*, who are often the most altruistic of all. Perhaps they have been encouraged by their employers, who want to demonstrate that their business is an active partner in the non-commercial life of the community; perhaps they want to help a local school but do not have the connections that other governors have. More likely, people in this category never really had any intention or aspiration to be a school governor until the telephone rang and a friend said that he or she had mentioned the governor's name for co-option – and accountants, architects, surveyors and fund-raisers are particularly vulnerable to this approach! Often, once such individuals become governors they demonstrate enormous commitment and energy. They are not used to failure in their professional lives and would not feel comfortable about being associated with failure in such an important voluntary role within their local community.

All of these categories of governor have a strong vested interest in the success of their school, but each brings to the task of governorship a distinct set of experience, skills and priorities, emanating from both their personal and their professional involvement.

Why is it then, with such a collection of self-motivated and influential individuals, that so many governing bodies find it hard to find a rhythm, get themselves organized and really feel in control?

Why is being a governor so difficult?

The answer to the foregoing question lies in the complexity of the role of the governing body and in its relationships with those it is there to govern, that is the headteacher and staff of the school.

The Education (School Government) Regulations 1989 defined the responsibilities of governing bodies in preparing them for Local Management of Schools. Subsequent regulations and the Education Act 1996 defined even greater levels of real power and influence, not to mention accountability, for governors in their schools. However, legislation has been deliberately vague about where the role of the headteacher stops and where the role of the governing body begins. Of course, governors' responsibilities are clearly laid out, in the Governors' Guide to the Law (published by the DfEE) – for example, there is an excellent table therein showing exactly what it is that governors are accountable for and what it is that the school's management team is accountable for. However, in trying to unpick these accountabilities there are many potential obstacles to both harmony and clarity in the relationships between governors and their managers.

Take, for example, the teaching staff. The governors have sole responsibility for appointing teachers; a headteacher cannot do this alone. Often a governors' personnel committee will have a delegated power to sit with the headteacher when appointments are made. Permission will have been granted by the governing body to make the appointment, and the results will be reported back at the next meeting. But although the governors have responsibility for appointing teaching staff, it is the headteacher, and only the headteacher, who has responsibility for deploying them. The governors may have appointed to a specific job description, such as a responsibility for music or the humanities, but the headteacher decides which class or classes they will teach.

In relation to the curriculum, the governors determine the curriculum to be taught and have to decide on a curriculum policy statement. The headteacher, however, decides how to implement that curriculum. Decisions such as those on setting and grouping would be professional decisions and not decisions made by the governing body.

Working with the headteacher

It is intended that governors and headteachers work out between them the most comfortable and practical way of working together. In governing-body meetings, however, there are often some tensions – mainly healthy tensions, thankfully – as governors and their headteacher explore the boundaries and intersections of their responsibilities. The best meetings are those where there is a free flow of information to governors on the school's performance, pupil profile and other indicators. In these meetings the governors, from a position of knowledge based on that information, make a point positively to question and probe before coming to an agreement or making a decision on an issue.

Whilst governors such as these are fortunate in having access to information, there are many other governing bodies who do not have access to the information they require in order to fulfil their responsibilities correctly. How many governing bodies by the end of their first autumn term meeting (if they have two meetings a term) have still not received a detailed report on pupils' performance in the national assessment tests sat in the previous May or June? How many governing bodies can describe to interested parents the quality-control techniques used by their managers to ensure good-quality teaching and learning in the classrooms? How many governors know how large the projected carry-over is in the current school budget and what the implications of that are for financing the school in the following year? How many governors have truly played a part in developing their own school's development plan? How many governors are able to identify what the main agreed priorities for their school are this year and how well they are progressing in meeting these priorities?

Is it fair to expect governors to know all these things? Is it fair to expect governors to know that they should know all these things? Even in the best local education authorities, governors have precious little training prior to taking on their responsibilities and not significantly more training afterwards. Governor training costs money and governing bodies are, correctly, quite selective about the training they purchase. Attendance at training, like attendance at governing-body meetings themselves, is voluntary and unpaid and is usually in the individual governor's own time.

Influence or information?

Headteachers, however, do know the answers to the above questions – or, at the very least, they have access to all this information and should be able to find the answers quickly. This is because everything that a school governor is responsible for on a volunteer basis automatically comes within the territory of a professional, paid headteacher. Information is power. With access to the appropriate information, obtained via the headteacher, a governor can interpret, influence, control and project. Without information, a governor can only listen to the interpretation of others. Without access to the original facts, a governor has no basis for challenging an interpretation and so can be (even unknowingly) influenced by the 'spin' that others put on the information . A governor's response can be controlled by the amount of information given and the way it is given. Future pathways and events can be forecast based only on what is heard from those with information rather than from what the governors should know as the accountable body.

Take, for example, the presentation of national assessment results to the governors. It is a long time between May and October, and once the summer holidays are over attention is usually being paid to the challenges of the new academic year rather than going over the events of the last. There have been governing bodies who have never received their school's results, because some headteachers forget to schedule them on the governors' meeting agenda and the governors themselves forget that they haven't seen them. There may be no denying the facts that in some years the results may be significantly lower than in previous years. Some governors may be given the results with a defensive commentary, excusing the lower-than-expected results, that focuses solely on issues such as:

- ❑ the high level of special needs children in that cohort;
- ❑ the cohort being generally less able than the one for the previous year;
- ❑ the test paper being unusually difficult and causing nationwide concern.

Governors can often look uncomfortable about this kind of informal commentary, but nobody wants to be the one who challenges because to do so could give the impression that the headteacher is making excuses. Even if governors may think they should be given more information or a better explanation, it takes a brave individual to be the first to say so. However, courage shouldn't come into it; governors are accountable for the success of the school and shouldn't be afraid of seeking information when it is not volunteered by the school.

Governors have a right to know the facts behind each statement. For instance, questions such as 'How many special needs children were in that class compared to the year before?' need asking. There will be more uncomfortable looks when the answer begins 'Well let's see, we had Harriet Thomas; what's his name ... Thomas, Thomas Dillon; and that really badly behaved child ... I'd have to check back over the records ...'. Decision time for the governors, who should challenge further and embarrass a totally unprepared and probably generally well-meaning headteacher, or else accept that this is the (poor) quality of the information received this year and make a mental note to address it before next year. However, it may come as no surprise to know that many governors make this mental note – and then make another the following year!

Governing bodies with good practice will have agreed with the headteacher in discussions about the conduct of governor meetings that all data relevant to all presentations would be shared in an appropriate form. So the presentation on lower-than-expected SAT results might then look like the following:

The high level of special needs children in the cohort. The number of children carrying statements of special educational need, together with the nature of that need, will be given for each of the classes in the assessment years (Year 2 and Year 6). In addition, the total number of children on the special educational needs register, at each level, will be published to the governors for each class.

The cohort being generally less able than the one for the previous year. The results for the end of Key Stage 1 will be shared, showing the proportion of children attaining at the average, above the average and below the average. The same results for the higher-scoring cohort in the previous year will also be shared. Both classes' results should show similar progress over time, but the lower-scoring class might well be proportionally lower than the other in both sets of data. Other data, such as reading ages, may also be used to demonstrate the point. Where pupils have moved into the school since the last Key Stage assessments, or have left the school before taking the tests, the effect of their grades will be taken into account when presenting the data. A similar analysis will be presented for the Year 6 cohort.

The test paper being unusually difficult and causing nationwide concern. The subject co-ordinator or assessment co-ordinator will demonstrate the difficulties in the test papers. References will be made to appropriate correspondence, information from other schools and the LEA to support this statement.

Open sharing of information by the headteacher is a sign of confidence. It means that the headteacher trusts the governors to understand the circumstances in which the school is working. It means that the headteacher is happy for the governors to work with the management team of the school in successfully addressing the issues that arise from these circumstances.

A successful primary school plummeted down the local league table in one year because of special-needs factors and pupil movement. The area in which this school is sited was at that time particularly turbulent, with many parents selecting the independent sector at the slightest hint of dissatisfaction with the local school. There was also a great deal of movement in and out of the area as young professional parents progressed up the career ladder.

The headteacher hid nothing. She presented the facts as they were to governors and parents alike. She applauded the successes of those children with learning difficulties who had progressed at least in line with expectations, as well as applauding the successes of the more able children. Information was shared fully, and the headteacher was able to demonstrate that her school served all its pupils well and not just the most able. Parents and governors were not concerned about the league-table position because they had the more salient facts.

There was no panic, no deception, no evasiveness – just the truth, well presented and clearly explained.

Summary

People become governors for a variety of reasons but each one believes that they can really help their school. Governors often underestimate the amount of time and commitment needed to fulfil the role properly. They have to know their school well to be of any use.

Governors need to develop a working partnership with the headteacher. Information must be freely available so that governors can come to their own conclusions, guided by the headteacher as necessary, about the health of the school.

THE NEED TO ASK QUESTIONS

If governors are to be effective in supporting their school and in moving it forward, they must feel fully informed and must be satisfied that relevant issues, be they to do with budget setting, school development, planning, academic achievement or buildings maintenance have been explored properly.

Sharing the workload

Governing bodies are often organized into working groups or subcommittees. The latter can take decisions on behalf of the full governing body, but only if the power to do so has been delegated to them. (There are, however, regulations determining what can and cannot be delegated to these committees, and these are spelt out in The Education (School Government) Regulations 1989, reg 25.2.) In contrast, working groups usually explore issues fully on behalf of the full governing body and report back with a summary of the issues and a recommendation.

Many schools have committees or groups covering curriculum, finance, personnel and premises, although variations on these can often be found – for example, finance and resources, personnel and pupils, and policy. When either of these systems work, they work extremely well. The enormous workload of the governing body is spread fairly between the groups and full governing body meetings are kept to a reasonably respectable length. However, in order for such arrangements to work, they require the presence of several key influences and factors.

Firstly, there is the chair of governors, who should lead the full governing body in the organizing of working groups and subcommittees and who should also ensure that the full governing body has clear expectations of each group. This involves making sure that each group has a clear brief or (in the case of a subcommittee) clear and agreed terms of reference. It means making sure that there is a convenor or chair for each group, and that meetings are scheduled in advance of full governing-body meetings. It means making sure that agendas are discussed with the headteacher and chair at the preparatory stage and that minutes are circulated after the meeting but before the full governing-body meeting. It means making sure that governors on each group have appropriate information relevant to the agenda and are prepared for the meetings. And it means trying to ensure that all governors turn up for meetings of the groups on which they sit.

The headteacher – the second particularly influential person within the governing body – should ensure that each group is serviced with relevant documentation and attended by appropriate professional staff. The headteacher should not be expected to, nor feel honour-bound to, attend every meeting of every working group; the load should be spread evenly between the head and deputy head and other responsible post holders for whom it may be appropriate to attend.

With such an arrangement as described, the governors should have a system that deals with an agreed agenda and that reports back to the full governing body. They should have the information they need to make decisions, and nothing should prevent them from making the best-informed decisions that they can – nothing, that is, except for their reluctance to dig deeper, to seek clarification and to ask for explanations or more detail. The best-organized governing bodies can sometimes perform short of expectations because they didn't ask enough questions at the right time.

Most headteachers start work sometime between 7.30 am and 8.00 am. They finish at school between 5.00 pm and 6.00 pm, and on the day of a governing body meeting they often work right through to the start of that meeting, perhaps to 7.30 pm. Many governors have worked similar hours, sometimes miles away from home, and if not in an office or factory then at home with children. Usually, everyone is tired at the start of a governing-body meeting and yet, out of courtesy probably borne out of being on the headteacher's own turf, it is often the headteacher who is acknowledged as being the most tired and hard-working individual there.

That may or may not be the case, but it is often the perception. And the perception leads to an unwillingness on the part of many governors to put the headteacher on the spot. They are often reluctant to ask for more information, and perhaps to express dissatisfaction with the quality of information that has been shared.

Knowing when to ask questions

Yet the asking of questions does not necessarily mean an absence of trust. On the contrary, it can just as easily be an acknowledgement of professional expertise and a recognition of the importance of the relevant aspect of the headteacher's work to the overall success of the school. Whether the issue in question be school finance, personnel, resourcing or attainment, the following questions are essential for governors to ask themselves before making a decision on a related issue:

- ❑ Do I understand the issue being discussed?
- ❑ Do I know what the likely outcomes could be at the end of the discussion?
- ❑ Do I understand the data or information being presented?
- ❑ Do I know that there is no other data or information I should be seeing in relation to this issue?
- ❑ Are the alternatives being fairly described and explored?
- ❑ Do we have to make a decision today, or could we request more information and for research to be carried out?

If there is any doubt in a governor's mind about the answers to the above questions, immediate clarification should be sought before the debate moves on. Volunteer non-specialists must not feel embarrassed about asking employed professionals for explanations. If it slows business down, then so be it: better to have a fully informed governing body making the right decisions, and being accountable for the right decisions, rather than an ill-informed one making risky decisions and being held accountable for those.

A previously very stable primary school lost several teachers in the same year that a new headteacher was appointed. The decision was made, on the advice of the new headteacher, to appoint three newly qualified staff because the school 'needed new blood and fresh ideas' and 'this option will also be good for the school budget'. Nobody asked how, with almost half of the class teachers being new to the profession, let alone new to the school, teaching standards could be maintained. Nobody asked who would undertake co-ordination of the subjects of the curriculum and aspects of school life that the departing members of staff had managed. Nobody asked whether the smooth running of the school would be disrupted by such a move and whether the quality of the children's education would be protected.

The decision was made, unchallenged by governors, who deferred to their new headteacher. Within two years, all three new teachers had left, feeling unsupported in some cases, and, remarkably, the decision to replace like with like, and appoint newly qualified teachers again, was accepted by the governors.

The school went into rapid decline – a decline that the presence of more experienced teachers and of more demanding governors could have prevented. For the sake of not wanting to challenge a new headteacher and not exploring the issue fully, the education of hundreds of children was put at risk. The payback time for this governing body came when they realized that they were being held to account for the failures of the school. Although all the governors were new to the school since the original decisions were made, they carried the accountabilities and consequences of the actions – or lack of actions – of their predecessors.

Governors who had looked uncomfortable at the time of the earlier decisions knew that something was wrong, but they had not had the courage or confidence to speak out at the time. Some may have made the decision not to be on the governing body when the consequences of these actions surfaced. Others may have had doubts about the headteacher's competence but, without a clear steer from other governors – and particularly the chair – they decided to let matters rest. Throughout all of this, the quality of children's education was being compromised.

What to do when trust breaks down

Governors need to be able to trust their headteachers. When that trust breaks down, either because information is not properly shared or because there are clear examples of poor management that should have been avoided, then governors must feel that they can take appropriate action. Indeed, the law now requires LEAs to have a strategy in place whereby governors are informed if there are serious concerns about the competence of a school manager. The governors then have a duty to act on these concerns and inform the LEA exactly how they will address them, giving a timescale within which action will be taken and improvements monitored.

Of course, a good governing body does not need information from an LEA telling them if things are going wrong. If they have been asking the right questions, they already know themselves and will be the lead partners, perhaps with the LEA, in addressing the issues before they become too serious.

Summary

Governors need to organize themselves so that the wide range of tasks and responsibilities can be logically and fairly shared across the whole governing body. Each working group or subcommittee needs to have access to the information it needs to carry out its responsibilities properly.

Governors should be clear, as meetings progress, that they fully understand the issues and can take informed decisions. The consequences of not probing issues fully enough or not getting a clear and rational explanation for a particular course of action can be significant.

THE NEED TO CHALLENGE ON EXAMINATION AND TEST RESULTS

Challenge does not have to be about confrontation. Indeed, healthy discussion leading to a full understanding of the issues can give reassurance to managers and teachers that the circumstances of their work are fully understood.

Pounds and pupil performance

The 1990s have seen enormous changes in the way in which governing bodies operate. The decade has also seen a significant shift away from an arms-length involvement in recruitment and policy rubber-stamping to an increasing awareness of what the circumstances and influences are that makes their school what it is. Two areas in particular have dominated governors' thinking.

Firstly, from the earliest days of the Local Management of Schools the emphasis of the responsibilities of governors was placed on financial management. Indeed, the earliest schemes were referred to as Local Financial Management of Schools. In the past 10 years many governing bodies have developed skills way beyond what they would have imagined necessary in the early days of reform as, with the help of local authority financial managers, they have begun first to walk and then to fly – almost unaided – in relation to the planning and management of their own school budgets.

The second area has been approached much more cautiously – by everybody, including government and school governors. Ten years ago national assessment tests were unheard-of in most schools and the publication of the results, in the form of league tables that would be taken seriously by parents as well as politicians, was something that many believed could never happen. One step at a time, successive governments have introduced and refined national assessment tests. Most teachers, having initially written them off as time-wasting and uninformative, now find them a useful referencing tool and a good way of verifying other kinds of judgements on children's attainment. Many parents find their publication and the decoding of local league tables vital in selecting schools and, in some areas, in challenging their children's schools. As the measuring of pupils' attainment continues to evolve, the socio-economic factors that many protested about because they were not present in the early days are now beginning to be taken into account.

Furthermore, schools are now being given data that allow comparison not only with local and national schools but with schools 'of a similar type'. This means that they can identify within that data schools that have a similar profile for, for example, the number of children on free school meals or the number of children who have English as an additional language. Indeed, when schools come to agree their literacy or numeracy targets with their local authorities, factors such as these can be taken into account. Although the published national assessment test scores do not currently provide this comparative data – and therein lies a significant fault with the system – the data are available in schools for them to use.

Making use of available data

So how do governors make use of this information to help them interpret their school's national assessment test scores or other public examination results? It is not a question of finding excuses to justify below-average performance; it is too easy and simplistic to look at a below-average set of test scores, look at another table showing above-average free school meals take-up, and breathe a sigh of relief by justifying one set against the other. A set of test scores that appear below the average needs much more careful analysis than that, starting with the school's own expectations for that cohort the year before they took the test and, in the case of national assessment tests, including an analysis of the school's Teacher Assessment scores. Any significant difference (and remember that a significant difference might be deemed to be 10 percentage points or more in a small primary school) between the school's forecasts or Teacher Assessments and the final scores should result in close scrutiny of internal assessment and teaching issues. For example, where teachers project that 84 per cent of children in a cohort will attain the national average and only 70 per cent do so, this could point to a number of other factors such as:

- ❑ inexperienced teachers or lack of training in assessment practices for teachers, leading to inaccurate forecasts in assessments;
- ❑ teachers with low levels of subject knowledge in the area being tested who, not understanding the subject fully themselves, will not be able to help children who have difficulty grasping key concepts and skills;
- ❑ a lack of previous, accurate assessment informing the teacher's teaching – if previous assessments have not been correct or if results have not been passed on, a teacher may not have sufficiently detailed knowledge of each individual in the class to set them appropriately challenging work.

Only when the governors are satisfied that the teaching and assessment activities themselves are secure should other factors be taken into account. This could then include the examination of key data such as:

- ❑ the class's previous test scores;
- ❑ the proportion of children with special needs in the class, their level on the special needs register, and the nature of their need;
- ❑ the proportion of children receiving free school meals;
- ❑ the proportion of children with English as an additional language;
- ❑ the proportion of refugee or traveller children;
- ❑ the level of authorized and unauthorized absences by pupils;
- ❑ the proportion of children, and their expected level of attainment, who have left the school since the last tests were administered;
- ❑ the proportion of children, and their last recorded level of attainment, who have joined the school since the last tests were administered;
- ❑ a comparison of results against schools with similar characteristics.

Such an exercise is not one that is particularly simple or that can be done at a full governing-body meeting. However, governors could expect that some of their members be involved with professional staff of the school in such an analysis, as a matter of course, before the results are shared with the full governing body.

Using understanding of data to influence practice

Knowing the results and having a feel for the factors that influenced them is still only half the story. Socio-economic data comparisons may be used to help explain why a school performs poorly, but the data cannot be used to justify poor performance. The point when a school feels confident that it knows why certain test scores or examination results have been weak is the point when the school should begin thinking about how to compensate for this weakness. A family taking free school meals, for example, does not always mean that the children live in unsupportive home circumstances, but it *can* be an indicator of stress at home where the emphasis is on the survival of the home or family rather than the detail of the child's education. In some cases it could also be an indicator of under-nourishment. (Many schools have considered a range of responses to these circumstances, including the provision of breakfast facilities, homework clubs and other forms of support and care to help these children.) Other factors, such as low levels of understanding in English, may cause a school to review its reliance on written English worksheets in lessons. This may result in other changes in teaching practices, such as more careful planning and the use of targeted language that is accessible to all pupils.

So, it is not only having the information that gives governors the power to influence change in their schools but knowing how to use that information. Governors who ask the questions that get to the truth behind the presented facts are serving their schools and the interests of the children well. Governors who accept an analysis unchallenged are not recognizing their responsibilities properly and are not helping or supporting the professional staff of the school as they should be doing. The perspective of governors in interpreting data, together with a collective local knowledge that surpasses probably even that of the headteacher, is too valuable to be ignored.

Where a governing body engages the professional staff and itself in informed discussion about pupils' performance data, it is strengthening the partnership that exists between the managers and themselves in their school and, by recognizing and understanding the issues behind the pupils' performance, is sharing the responsibility with the school staff for addressing those issues. In contrast, a governing body that simply accepts what is, or is not, given to them is failing to support its teachers and its pupils in recognizing that standards can be improved and in helping identify the strategies to enable this improvement. The school with a governing body that questions and reacts may, if this is done positively, be on the road to improvement, whereas the school with a governing body that does not challenge is almost always on a path of slow and steady decline.

Summary

Governors have in the main acquired a good financial understanding of their schools. This now needs matching by an understanding of what factors influence pupil performance in their school. Having this understanding can help produce the strategies that will raise performance.

THE NEED TO VISIT THE SCHOOL

Years ago, the image of the school governors walking into the classroom was usually portrayed by a smartly dressed, often overweight, individual with an unmistakable look of self-importance, accompanying a rather strict looking headmistress into a classroom. There, a cowed teacher and even more cowed pupils waited in anticipation for a nod of approval from the Very Important Visitor. It is unlikely that, if the nod came, slates and dusters would be thrown into the air together with shouts of 'The man from the guv'ners, he say "Yes!".' However, the sentiment was probably just the same, such was the awe these people held over those they governed.

Getting involved within the school

Today, a school governor is much more likely to be seen in a classroom helping children read or engaged in some other support activity on a rota with other parents. Governors who visit their schools often do so either in this sort of capacity or for meetings, for many governors who spend time in classrooms are either parents or retired governors with time to spare. Other governors who are working may simply not be available for this kind of activity on a regular basis.

Yet it is important that each governor should visit the school, in the capacity of governor, periodically whilst the school is in session. It is important for the teachers to see who the governors are and to see that they are interested. It is important for pupils to begin to understand, as part of their education in citizenship, that schools are a part of the local community and that people from the community give up their own valuable time to help the schools run smoothly. It is important for governors to see first hand the children learning and playing together and to feel first hand the ethos and character of the school for which they are responsible. Yet there are governors who, because of their working commitments or because organization of these visits has never really been addressed, have never visited their school when it is in normal session.

Some schools and governing bodies go to great lengths to try to rectify this situation. Schemes are implemented such as 'governor of the week' or 'governors of the month', where governors are expected to visit to meet staff and pupils and, over a longer period, be involved in some sort of project. Another scheme is the 'attached governor', where each governor is attached to a particular class, with the expectation that they will visit and 'get to know the children and the work they are doing'. Rarely do such schemes work successfully. Usually, the same governors – those for whom finding the time is easy – respond well, but the usual pressures of work in other governors' professional lives mean that they have difficulty keeping up the commitment. Many schools have tried such schemes and quietly let them die, accepting that no more can be expected of hard-working volunteers. Whilst they find it odd that there can be school governors who never see the school in operation on a normal working day, they accept it as an unavoidable fact of modern-day life.

There is, however, no need to accept this sorry state of affairs. Busy as many governors are, there is no reason why even the most demanding of employers should not agree to one or two hours occasionally to allow governors to fulfil their duties properly. The main impediment to many governors not visiting their schools more regularly is not being unable to get the time away from work or home but being unable to justify the visit: whereas hearing a child read may be second nature to some parents (and particularly those that are interested enough to become school governors), it is quite a daunting experience to others.

Focusing the visits

Schools and governing bodies should be very clear about why governors should visit their schools and what form those visits should take. It is important that, together with the factual data about their schools and pupils, governors have the opportunity to see what that actually means for the school in practice. For example:

- ❑ What difference does a large proportion of non-English speakers have on a class?
- ❑ What do a teacher and school have to do to address this?
- ❑ What difference does a high proportion of pupils on the school's special needs register have on a class?
- ❑ How does the school and the teacher support special needs pupils?

Whereas it would be inappropriate for the whole governing body to be taken round the school *en masse*, or even in large groups, it is perfectly practical to arrange for one or two governors at a time to be given a tour. Their visit could focus on certain achievements by the school's pupils and certain issues that the school is having to address on a day-to-day basis. In this way, the governors get the opportunity, at least once a year, to place the data, policies and initiatives that they are discussing at their meetings into the context of the pupils and teachers working in their school. Although the ideal would be one visit a term, even one visit a year will help make more sense of some of the debates and discussions that the governors are undertaking.

Such visits need careful planning, however. They need to be scheduled well in advance – perhaps up to a term. They need to be carefully balanced so that the individual governors, if the visit is not concentrating on a specific aspect such as those highlighted by the questions above, are not seeing an unrepresentative sample of the school's work and are not gaining a distorted perception of the school's problems and successes.

Many schools adopt the good practice of sharing curriculum presentations with school governors. A subject co-ordinator, supported with samples of pupils' work, can explain in outline to governors what material is taught, the challenges that the pupils and teachers face in this subject, and the strategies used to overcome them. This practice helps further governors' understanding of the work of the school.

One word of caution is necessary here. When looking at pupils' work or when visiting a classroom, a governor is not there to make a judgement but to observe. It may be appropriate to ask questions of the headteacher or teacher afterwards, but governors should avoid coming to conclusions themselves about the quality of what they see in a single visit. Governors always come away with a good – or sometimes not too good – impression of what they have seen, and it is important to share this impression with the headteacher. The headteacher can do much to pre-empt difficult post-visit exchanges by explaining, before the visit, what the governor is likely to see and how representative (or otherwise) this is likely to be of normal day-to-day life in the school.

Summary

It is essential for every governor to visit the school periodically whilst it is in session, although the difficulties that some governors face in doing this are acknowledged. It should be clear what the focus of the visit is: is it a general walk round, or is there a specific issue to be examined and discussed later?

Governors should visit classrooms to deepen their understanding of the work of the school and not to make judgements. However, if questions or concerns have been raised in the minds of governors about what has been seen on a tour of the school, they should be raised with the headteacher directly.

THE IMPORTANCE OF GOOD MEETING STRUCTURE

The overall organization of the governing body was discussed earlier in this chapter. Now it is appropriate to look at how the governing body, and its subcommittees or working groups, organize their own work.

Meetings and chairs

Most governing bodies meet around six times a year unless there are special circumstances that require additional meetings. In addition to their scheduled meetings, there are the working groups or subcommittee meetings, often held before each full governing-body meeting. There is also the annual parents' meeting, which is the governors' opportunity to explain the business of the previous 12 months and the parents' opportunity to question the governors on issues of interest. Like the annual report of governors, which is considered at the annual meeting, the annual parents' meeting has by law to have a specific content.

Governors meetings themselves are less consistently structured, and this can lead to disproportionate amounts of time being spent on relatively minor issues, while more important issues get squeezed into the later sections of a meeting and suffer from insufficient time for proper consideration. Some governors' meetings are notoriously unstructured and last far too long, with the result that important issues are discussed when everybody is feeling tired and restless. With most governors attending at least 12 meetings year, it is important to give thought to how the meetings are structured and conducted.

Good chairing helps enormously. Chairing a meeting well is a skill that not everybody possesses. A good chair can steer a governing body through the most demanding of agendas without the meeting lasting more that a couple of hours and still leave governors feeling satisfied that they have considered all the issues and have made well-founded decisions. A good chair paces a meeting properly and ensures that the minor items are quickly dealt with, having been given proper attention. A good chair ensures that all governors have their say and will make sure that every governor has an opportunity to speak when important decisions have to be made. A good chair does not dominate the meeting and will make sure that nobody else does: discussions are kept on track and will not be allowed to wander or deviate; anecdotes are kept to a minimum.

A good chair is mindful that a governing body works at its best when there is consensus or when the minority view has been fairly presented and discussed. Decisions made by the governing body are not the decisions of individual governors nor, once taken, are they any longer the views of individual governors. Once a decision has been made, even if it is a majority view and not unanimous, it becomes a decision or a view of the whole governing body. A good chair of governors is mindful of this when conducting business and will ensure that individual governors also understand this.

Nevertheless, good chairing is not enough on its own to make a successful governors' meeting. Unless thought has gone into the agenda and what business will be discussed at each meeting, meetings will still lack focus and direction.

Planning the long-term agenda

It is worth spending some time giving a brief outline of the more important issues that have to be considered throughout the year. Whilst an academic year starts in September, for many governors the start of their year is the financial year starting in April. Governors meetings work on a cycle and the beginning of that cycle is going to be, for most, consideration of the school's priorities and budget setting. This process, often taking place in the spring term, usually involves a thorough review of the school development plan, evaluating progress made on the previous year's targets and setting or agreeing the following year's targets. The plan should be properly costed and priorities should be agreed by the full governing body with consideration as to the financial resources available to them. As part of a review of the previous year's progress, it is good practice to review the previous year's academic results and other performance data.

A school's budget will be drafted in the spring term and probably agreed at the first meeting of the new financial year in the summer term. In the spring and summer terms, there will be consideration of the curriculum and any changes that may need making for the coming academic year, including consideration of staffing structures and changes in personnel. Throughout the year, governors will be considering buildings maintenance and urgent resourcing issues and will also be responding to deadlines and initiatives from their local education authority or central government. Every policy that the governing body and school have should also be scheduled for regular review – a period of, say, three years is appropriate for most policies.

This description gives some idea of what the annual workload of a governing body is like. The description is not comprehensive and can also be significantly affected by local circumstances or local priorities.

Controlling the short-term agenda

When considering a workload such as just described, it is easy to see why good governing bodies make extensive use of their working groups or subcommittees. In any school, there are always important issues or initiatives relevant to the school that need to be considered by the governing body. There may, for instance, be a need to resurface a seriously pitted play area and a need to provide additional cloakroom facilities in one part of the school. Both have health and safety implications and both could be equally costly. The governors may well be unable to afford both in the short term, and could easily spend an hour or more at a full meeting discussing which one should be tackled first and how it should be funded. It is unlikely, however, that this sort of debate, involving the whole governing body from a 'cold start', would be useful in finding the best outcome, and this is where working groups or subcommittees come into their own.

A well-organized governing body will, after a brief discussion, delegate to one of its working groups the responsibility for researching the problem, costing potential solutions and identifying possible sources of additional funding. Some governors may decide to put specific project groups together to work solely on one-off issues of the type given here; others may have standing committees ready to tackle projects of this kind. The delegated working group will then be asked to present a detailed report and recommendations at the next meeting of the governing body. Twenty minutes or perhaps half an hour can then be spent in focused discussion of the solutions rather than in unfocused discussion of the problems.

The main governing body meetings can operate using a simple agenda such as that set out below.

Governors' meeting: typical agenda

1. Apologies for absence.
2. Minutes of the previous meeting.
3. Matters arising from these minutes if not covered in later items.
4. Report of Curriculum/Policy Subcommittee.
5. Report of Staffing/Personnel Subcommittee.
6. Report of Finance/Resources Subcommittee.
7. Special item (eg consideration of an LEA officer's report).
8. Any other business (only business previously notified to the chair before the start of the meeting will be considered).

It is very important to control the final item (any other business) quite strictly. Many a well-paced and potentially brief and focused meeting has been destroyed through the chair allowing an unfocused diversion into issues of dubious relevance brought up by governors with a point to make under AOB!

Attention should also be paid as to how the agendas of subcommittees or working groups are drawn up. It is not sufficient to have the structure in place and expect each group to generate its own priorities. A common agenda structure should be agreed and, at the start of each year, core agenda items relating to the cycle of a governing body's work can be scheduled. By using this strategy the convenor or chair of a working group can, by agreement with the chair of governors, control the subcommittee meetings. Important items can be dealt with in time to feed back and inform a full meeting of the governors. New items, emanating from a previous governing-body meeting or from correspondence via the headteacher and chair, can be dealt with similarly.

A cycle of subcommittee meetings, each taking place before a full governing-body meeting could take the form outlined in Figures 5.1–5.3 for each term of the academic year respectively. Before each meeting there would meetings between the convenor and the chair of governors and headteacher. After each meeting there would be appropriate contact with the convenor of other subcommittees. It makes sense to have the Finance/Resources Subcommittee meeting as the last in each cycle because the financial implications of the business conducted at the other meetings will have to be considered before that business is discussed by the full governing body.

Meeting	*Subcommittee/group*	*Standing items*
Autumn Term Meeting 1	**Curriculum/Policy**	Review policies scheduled for review this term
		Consider appropriate items from the school development plan
		Consider new policies
	Staffing/Personnel	Review induction arrangements for new staff
		Review staff appraisal arrangements for the year
		Consider appropriate items from the school development plan
	Finance/Resources	Consider budget position to date
		Review arrangements for subject/departmental resource allocation
		Consider appropriate items from the school development plan
Autumn Term Meeting 2	**Curriculum/Policy**	Review policies scheduled for review this term
		Consider appropriate items from the school development plan
		Consider new policies
		Consider business delegated from full governing-body meeting 1
	Staffing/Personnel	Consider resignations and draw up job descriptions and person specifications for positions to be filled
		Consider recruitment strategies
		Form appointment panels
		Consider appropriate items from the school development plan
		Consider business delegated from full governing-body meeting 1
	Finance/Resources	Consider budget position to date
		Consider financial implications of work of other committees
		Consider appropriate items from the school development plan
		Consider business delegated from full governing-body meeting 1

Figure 5.1 *Subcommittee meeting cycle (autumn term)*

Meeting	*Subcommittee/group*	*Standing items*
Spring Term Meeting 1	**Curriculum/Policy**	Review policies scheduled for review this term
		Consider new policies
		Consider business delegated from full governing-body meeting 2
		Review of progress on School Development Plan and consideration of new priorities
	Staffing/Personnel	Consider business delegated from full governing-body meeting 2
		Review of progress on School Development Plan and consideration of new priorities
	Finance/Resources	Consider budget position to date
		Consider financial implications of work of other committees
		Review of progress on School Development Plan and consideration of new priorities
		Review of progress on School Development Plan and consideration of new priorities
		Business delegated from full governing-body meeting 2
Spring Term Meeting 2	**Curriculum/Policy**	Review policies scheduled for review this term
		Consider new policies
		Consider business delegated from full governing-body meeting 3
		Review of progress on School Development Plan and consideration of new priorities
	Staffing/Personnel	Consider resignations and draw up job descriptions and person specifications for positions to be filled
		Consider recruitment strategies
		Form appointment panels
		Review of progress on School Development Plan and consideration of new priorities
		Consider business delegated from full governing-body meeting 3
	Finance/Resources	Consider budget position to date
		Consider financial implications of work of other committees
		Review of progress on School Development Plan and consideration of new priorities
		Draw up draft budget for next year
		Consider business delegated from full governing-body meeting 3

Figure 5.2 *Subcommittee meeting cycle (spring term)*

Meeting	*Subcommittee/group*	*Standing items*
Summer Term Meeting 1	**Curriculum/Policy**	Review policies scheduled for review this term
		Consider new policies
		Consider business delegated from full governing-body meeting 4
		Consider progress on new School Development Plan and priorities
	Staffing/Personnel	Consider business delegated from full governing-body meeting 4
		Review of progress on School Development Plan and consideration of new priorities
	Finance/Resources	Consider budget position to date
		Consider financial implications of work of other committees
		Consider progress on new School Development Plan priorities
		Consider business delegated from full governing-body meeting 4
Summer Term Meeting 2	**Curriculum/Policy**	Review policies scheduled for review this term
		Consider new policies
		Consider business delegated from full governing-body meeting 5
		Consider progress on School Development Plan priorities
		Consider content of annual report to parents to be produced by group
	Staffing/Personnel	Consider resignations and draw up job descriptions and person specifications for positions to be filled
		Consider recruitment strategies
		Form appointment panels
		Review of progress on School Development Plan priorities
		Consider business delegated from full governing-body meeting 5
		Consider content of annual report to parents to be produced by group
	Finance/Resources	Consider budget position to date
		Consider financial implications of work of other committees
		Consideration of progress on new School Development Plan
		Consider business delegated from full governing-body meeting 5
		Consider content of annual report to parents to be produced by group

Figure 5.3 *Subcommittee meeting cycle (summer term)*

Governing bodies that adopt such structuring have a much better chance of spreading the workload evenly throughout the year and are not likely to be caught unawares by several converging high priorities all requiring immediate attention. Conversely, in less successful schools, discussion of many important items is constrained or missed because the organization of the work of the governors has not been considered and has not been appropriately planned. These are often the schools where governors feel less informed and less in control than their more successful counterparts.

Headteachers, too, benefit from a highly structured governing body. After all, if essential work is not done by the governors, it is often done by the headteacher or by other professional staff of the school. Such organizational systems make life easier for everybody, particularly when combined with good chairing skills. Governors will see how the cycle of their work progresses and the school will benefit from their increased knowledge of the school's circumstances and the current issues facing the school, its management and its governing body.

Summary

The work of governors is for the most part cyclical. Understanding the cycle and planning in order to spread the work over a full year is essential if governors are not to feel overloaded at times. Good use of subcommittees and working groups and good organization of meetings helps to keep all meetings efficient, focused and of reasonable length.

THE IMPORTANCE OF GOOD TRAINING

Nobody can be expected to undertake a role as demanding as that of school governor without being offered training. Yet training is not always at the top of governors' thinking when new members join the board.

Training as a priority

Remember the reference near the start of this chapter to accountants and surveyors being telephoned by well-meaning friends and cajoled or coaxed into sitting on their local school's governing body? Remember also the diversity of backgrounds of individual governors and the different types of governor on a governing body? All governors are subject to election, re-election or reappointment, dependent on their category and the type of school, every two, three or four years. Once the children of parent-governors leave a school, the parents are not eligible to stand again as parent-governors once their term of office is up. Teacher-governors of course must resign their position as governors if they leave a school to take up a post in another.

Unsurprisingly, there is a lot of change constantly in progress in a governing body. Most governing bodies have at least one change of membership every year, while some have up to half their number changing in some years. Sometimes there can be a combination of circumstances which sees a majority of governors being replaced in the space of one academic year.

Given these circumstances, there is obviously a need for somebody to have a clear eye on the training needs of the governing body, both as a whole and individually. Training is often essential for governors who are going to take part in staff appointments. Training should also be deemed essential for governors with specific responsibilities, such as for special education needs or for exclusions. Governors with responsibilities on a school's finance subcommittee would find training extremely valuable because LEA delegated budgets and financial systems often have significant differences to the budgeting and financial systems experienced in industry: not appreciating and understanding these differences can lead to high levels of frustration, often resulting in pointless criticism of local authority methodologies. Life is easier for everybody if governors from the world of commerce accept that they are working in a local government environment and not a business environment and furnish themselves with the appropriate knowledge. In most instances, the local practices are not 'worse' than those experienced in other environments; they are just different because of the way public funding is managed by central and local government.

There are training courses for governors offered by local authorities, diocesan boards, colleges and other providers, and these courses should cover every area that governors will need to know about. It is unrealistic, though, to expect all governors to attend all training, even all training that is pertinent to the subcommittee or group they sit on. Although governors can claim reasonable expenses from the school's budget, few do so and there is no other financial compensation for their attendance. Training is in their own time, in addition to the time they spend at governors' meetings, subcommittee meetings and visiting the school.

Good practice worth considering is for the governors to decide a minimum level of training for each new governor, plus a minimum level of training for governors sitting on different subcommittees or groups or with named responsibilities. This strategy makes more sense given that governor training usually has to be paid for by the individual. It is also worth remembering that new central or local government initiatives may also have implications for governor training and training budgets. Some governing bodies adopt the practice of having the equivalent of a training manager, where this individual keeps records of all training undertaken by individual governors and arranges training for new governors and for others when necessary.

The consequences of ignoring training

A primary school did not have any strategy for governors' training. It did not subscribe to any training provider's services and had no system of induction for new governors. The governors were very well-meaning individuals and this was proved by their very commitment to the school through taking office in the first place. However, they had little clue about their priorities in education.

The headteacher struggled. The difficulties faced in managing the school on a day-to-day basis were compounded by the governors not fulfilling their role as the demands of Local Management of Schools and the newly-introduced National Curriculum built up. Besides running the school, the headteacher ended up doing all the things that the governors should have done or, at the very least, contributed to. Problems were further compounded because of an overall shortage of funds, insufficient administrative support, and a deputy headteacher with a full teaching load. The school development plan was written in its entirety by the head; essential policies were written solely by the head; the annual report to parents was written mainly by the head; and the budget was prepared and monitored by the head.

The governors met in a totally unstructured meeting once or twice a term and rubber-stamped the headteacher's work with little meaningful debate and little evidence of understanding the issues being presented to them. They gave credit to the head for doing the work and looked a little embarrassed when it was mooted that they might have done a little bit more. Attempts to set up working groups failed because the only person driving them was the headteacher, and in the end it was quicker for the head to undertake the work of the working groups than to spend time trying to convene meetings.

> Little wonder, then, that the headteacher began to lose contact with the day-by-day running of the school. Stressed and clearly overworked, the frustration sometimes showed in the headteachers' dealings with teachers and parents. Governors were concerned about this and initially gave much more sympathy but unfortunately precious little else in the way of practical support. It eventually became clear that the headteacher was not managing either the workload or the school and as a consequence was falling out of favour with the staff and parent body.

This account graphically illustrates what can happen when a governing body fails to function. In this instance, the decline in popularity of the headteacher and school in the eyes of parents was the catalyst for the governing body to sort itself. However, instead of recognizing that one of the basic problems was its own lack of structure and its failure to carry out its role properly, the governing body began to challenge, in a repeatedly co-ordinated manner, the 'failings' of the headteacher. The headteacher eventually resigned and the governing body reorganized itself to conduct its meetings in a much more structured manner.

It is doubtful that the true cost of the earlier lack of good practice amongst the governors will ever be appreciated or fully understood. Had they been aware of their responsibilities enough to undertake the basic training offered to them it is possible that the governors of the school would have functioned in a much more structured and effective way. Through becoming involved in such things as development planning or setting the school budget – essential activities for the fulfilment of their responsibilities – they could have significantly eased the headteacher's workload, given much needed support, and taken part in informed discussions and decision making on the work of the school. It is possible that the headteacher, who would not have been working in almost total isolation, may have had a very different and much more positive experience at the school, with subsequent benefits to all concerned.

Summary

The responsibilities of governors are wide-ranging and complex. It is unreasonable to expect that they can be carried out without training. The consequences of governors not fully understanding their own role, or not fulfilling it properly, can be very damaging indeed.

CHAPTER SUMMARY

The main points from this chapter on governors are as follows:

- ❑ Governors are probably the most accountable people in the education process but are often the least knowledgeable about what is going on in their schools, especially in weak schools.
- ❑ A feature of insecure management is the economy with which it shares information. Conversely, a confident headteacher will share problems with governors as well as successes and will include governors in the decision-making processes to find solutions to these problems.
- ❑ Where governors do not have information, they cannot be included in these processes even where they have a statutory obligation to be so included; neither can they respond to parents when they are approached on these issues away from governors' meetings. In time, governors in schools where information is not made freely available will feel devalued and disenfranchised. One response may be to allow the headteacher to continue with a 'maverick' management style and hope that they are not in office if and when the governors are finally called to account. Another response, increasingly likely with forthcoming legislation on education, is that governors will see a conflict between themselves and their headteacher that can only be resolved through disciplinary action.
- ❑ Governors in weak schools do not pre-empt difficult situations by ensuring that they take responsibility for being informed.
- ❑ Governors who do not ask regular and planned questions on school performance are not fulfilling their responsibilities; neither are governors who do not challenge the school on the results it achieves. It is not enough to accept a headteacher's own analysis of results as final; statements made by the head should be tested and explored until the governors are satisfied that the conclusions reached are accurate.
- ❑ The 'interventionist' strategy is not about making life difficult for headteachers or about conflict. It is about accountable individuals ensuring that the professionals that they employ have considered all the performance data they have in the most appropriate way possible.
- ❑ Where such discussions occur, those discussions will strengthen good partnerships and will only put a strain on partnerships where there is not openness and honesty in the quality and interpretation of information that is being shared.
- ❑ Governors need to be aware of their duties and responsibilities under the law, and to be adequately trained to carry out those duties.

- ❑ A poorly organized governing body with poorly structured meetings, few opportunities for governors to visit the school, and an inadequate programme of appropriate governor training is less likely to succeed than one that addresses all these issues properly.

Part 2

Prevention and cure: knowing what to do and when to do it

6

A SYSTEM FOR KNOWING ABOUT YOUR SCHOOL

It is clear that successful schools are not created by luck or by circumstance. It is true that a particularly dynamic headteacher can compensate, through his or her personal skill and energy, for many deficiencies in a school. People tend to overlook even the most glaring of deficiencies if a headteacher is popular and the results in English and mathematics are good. However, most schools can improve if all parties concerned, including governors, the management team, teachers and parents have a clear knowledge and understanding about their own role and how they can contribute most effectively to building a successful school. Much of that contribution concerns the flow of information from one group to another, from individual to individual and from school to community. Much of that contribution comes from using the wealth of knowledge and expertise possessed by different groups and individuals within the school community.

WHAT GOVERNORS NEED TO KNOW

For governors, the need to know specific information has to be considered collectively. It is unrealistic to expect that each individual governor should be knowledgeable on most areas of the school's operations. It is more reasonable to expect individual governors to specialize through their working groups or subcommittees, and within those groups to develop specific knowledge and expertise. In this section the range of knowledge that a governing body should possess collectively will be considered together with why each area of knowledge is important.

Personnel matters

Every school should appoint staff within a staffing structure devised to meet the needs of the school. The staffing structure should be reviewed whenever changes on the curriculum or the school's priorities change, and that structure must be referred to whenever the school is planning to fill a vacancy. The staffing structure will reflect the school's priorities and will cover all professional staff, including management and support staff who work in (or service the work in) classrooms. Within the structure, positions of responsibility are identified and graded, ensuring that the total cost of the staffing structure falls within budgetary expectations.

The grading of posts must be done openly, and the rationale behind grading one position at a higher salary than another should be made known to all likely to be concerned or affected. Take the following example.

> A school made a decision to pay a teacher on a higher grade simply because of her previous experience rather than her new responsibilities. The teacher concerned was new to the school and her actual worth was unknown. Similarly experienced teachers doing a job with slightly more responsibility found themselves earning less than the new teacher. To make matters worse, staff were not all told the grade at which the new teacher was starting and found out through the grapevine.

The foregoing example was hardly a recipe for harmonious staff relationships! There could be an argument to state that it is nobody's business but the individual's and the management's when it comes to the level of pay for a particular job. However, in education we are still at the very beginning of individually negotiated pay packages, and while teachers are paid on a common pay spine with a reasonable expectation of a certain number of discretionary points for certain additional responsibilities, teachers will always claim a right to know how the salary bill in the school is shared out.

Paying the best to teach, not to manage

There are many teachers in UK schools who would gladly remain in the classroom with no management or organizational responsibilities if it were financially attractive to do so. The government's initiatives to make the job of teaching financially attractive to very able young graduates has therefore to be welcomed. However, there are very many able teachers already in the profession who must not be overlooked by this initiative. Good teachers can only function effectively if there is good management and, for those already in the profession, weak management will damage their chances of success or recognition.

There is little doubt that as new pay structures for teachers are introduced and governors begin to exercise flexibility in annual pay reviews, there will be enormous potential for perceived injustice and internal staff conflicts, which could be very damaging to staff morale and teamwork. In some schools this could actually affect standards, because all teachers are responsible for improving progress and attainment and not just a handful of identified high-flyers. The rationale for treating some teachers differently needs building into a carefully structured framework that recognizes the capacity for contributing to a team and middle management responsibilities as well as an outstanding ability (beyond that expected from all teachers) to motivate children and improve standards of attainment. All teachers, particularly the experienced and effective hands that have borne the brunt of years of upheaval and reform yet have stayed loyal to the profession and the children they teach, must be shown that any new salary structures are fair.

However, this should not mean that all teachers should benefit equally from salary restructuring simply on the basis of experience. One of the most frustrating and demotivating factors for many good teachers is that no matter how hard they work or how effective they are in their work, they may still end up taking the same salary home as the weakest and least effective teachers in their school. This is one reason why many teachers leave, or consider leaving, the profession. So why is there such a fuss about linking appraisal to pay? Is it not illogical to divorce a system that recognizes quality in an individual's work from reward within the salary structure? In the final analysis, few would argue that a system that values financially the weakest in the profession with the best in the profession is a system worth hanging on to.

Aiding good performance

Governors and managers need to develop a salary structure that supports a staffing model for the school, in terms of organization and responsibilities, that has the capacity to reward excellence in teaching within that framework. The staffing structure should be supported with individual job descriptions for each position. The job descriptions may have a generic core, but each set of responsibilities beyond this core must be clearly laid out for the post holder.

In some cases, the most recent job descriptions to be drawn up have been those of the headteacher and deputy headteacher. Friction can arise during the construction of even these job descriptions when there is an obligation for the school to produce certain key documentation: the deputy may feel that he or she has a full enough workload anyway, and the head may feel that the job ought to be delegated – so who does the tasks under consideration? (In one instance encountered, the task being argued over was the actual drafting of job descriptions for the headteacher and deputy headteacher – a task that neither of them should have been leading on in any case: governors please note.)

Further, governors need to know that suitable arrangements are in place for staff appraisals, because appraisals will be a useful tool in assessing the performance of individual teachers when considering discretionary pay awards. Appraisal must include non-teaching staff as well as teaching staff. In addition to statutory appraisal, or as an extension of it, many schools are developing, or have developed, a cycle of termly professional interviews carried out by senior staff. These are linked to observations of the teachers' work and result in the agreement, by both parties, of targets for the staff member being interviewed.

It is also good practice for governors to be regularly informed of the training undertaken by teaching and support staff. The school's professional development co-ordinator should be given the opportunity to explain to governors how the school's training needs are met, either through written reports or through occasional presentations, perhaps to the Personnel Subcommittee. Measuring the effectiveness of training in terms of faster progress or higher attainment for pupils is difficult, but consideration needs to be given, where possible, to the cost-effectiveness of training being provided.

Where governors are uncertain as to the way a school's staffing is structured and where they have not seen job descriptions or do not know the arrangements for appraisal or training of staff, their schools are likely to have personnel-related deficiencies that could impact negatively on the school's overall success.

Financial awareness

Ask governors what the most important job is that a school governing body has to do and the chances are that they will say that it is managing their school's budget. There is usually one individual who drives the budget-setting process. It is often a governor with appropriate skills, but in many schools it is the headteacher or a member of the management team. The importance is not so much in which individual holds the responsibility for budget development but in how well that individual leads the other governors – and particularly those involved in the Finance Subcommittee – in developing the budget and their understanding of the budget together.

Having a clear grasp of the financial circumstances of a school should not be the domain of one person alone. If it is the headteacher, governors will never feel empowered; if it is a single governor, the school may face difficulties when that individual's term of office is up.

Because of the way local government finance works, it is difficult to project a school budget accurately beyond a year or so. However, governors should all be aware of the overall state of the budget, the likely carry-forward at the end of the financial year, and the priorities being set, in addition to staffing and other general running costs, that the budget is funding in any one year. Regular monitoring of school expenditure, the evaluation of expenditure against pupil performance, and detailed financial reports to the full governing body every time it meets are all hallmarks of a school in control of its finances.

Accommodation and resources

The biggest single potential liability that a governing body has is the condition of its buildings, and the biggest demands on what is left of the budget after staffing costs and day-to-day running costs have been taken out are for major resources. These include such things as computers, specialist equipment, library books and CD ROM resources.

Although a local authority or other authority may be responsible for much of the cost of new buildings or major alterations, there may be minor building works that the school feels it desperately needs and that are low on the list of priorities of those other authorities. In addition, there will undoubtedly be maintenance works from time to time that are the responsibility of the school. It is not unusual for schools to make minor alterations and improvements themselves with funding raised by parents or through careful savings on the budget over a number of years. Sometimes schools can agree an element of shared cost with another funding body when planning more costly improvements. In any case, the school is usually solely responsible for the general maintenance and decoration of its own buildings.

Good practice in accommodation management involves a thorough examination of the school's site, its condition, and the priorities for the future. There should be an element of vision in this practice rather than simply planning to maintain the status quo.

A primary school drew up plans for significant improvement and development of the accommodation on its site. The plans included the replacement of 'temporary' classrooms with permanent buildings; the planning of modern, attractive and secure reception

> and office areas; and the remodelling of staff-room facilities. At the time of the creation of these plans, no one knew how the work would be funded; nevertheless, because the school had a clear plan it was able to take advantage of 'windfall' opportunities at the end of the next two financial years, and as a result some of the planned work was completed within three years.

The chances of that happening at all, let alone within the context of harmonization with possible future developments, would have been minimal without a clear plan. If plans are created that are able to be implemented in stages then, as funds become available, work can begin. Often, money has to be bid for quickly and the unprepared lose out. It is time well spent having a buildings development plan. After all, without a vision of where you want to go, the chances are that you will stay standing still.

Similarly it is important for governors in the appropriate subcommittee to know the condition, age and sufficiency of their school's major resources such as computers, library stock or other costly equipment that has limited life expectancy. Good forward planning makes the provision of a contingency for the gradual replacement of such resources possible. The alternative is being faced either with enormous bills for the complete replacement of obsolete equipment or having to run with old stock while newer equipment is gradually purchased. Governors that have a feel for the potential resourcing liabilities that they have in the school are going to be much better able to keep their school's resourcing up to date and sufficient, and this could contribute positively to the standards achieved by pupils.

Knowing the pupils

Governors in most schools would claim that they know the social circumstances of the majority of pupils and that they have an idea of the pupils' abilities on entry to the school. However, many are surprised when they see data and other evidence that disagrees with their perceptions. For example, a good number of governors underestimate the number of pupils receiving free school meals, while others are surprised when they learn of the number of pupils on the school's special educational needs register or the number of pupils who have English as an additional language.

If governors are going to have an appreciation as to how successful (or otherwise) their school is, they need a thorough understanding of the circumstances and abilities of the children that the school is dealing with. This is becoming increasingly important as governors move into the practice of agreeing performance targets for their schools and for subjects of the curriculum. If targets are to be attainable, they have to be realistic and within the capabilities of the cohort. The targets also have to be sufficiently challenging to move the cohort forward and not be too easily attained. Knowledge of not only the overall picture of pupils' backgrounds but the picture of each cohort as it enters the school will help governors and their managers identify trends and enable them to plan for meeting changing circumstances accordingly.

School development and policy planning

Ask most education officers what the single most important job is that a governing body has to undertake and the chances are that they will say it is forward planning (or school development planning). There isn't a difference in priorities here between education officers and governors, the latter being likely to cite financial management as the single most important job; it is more a difference of nuance. Education officials would, correctly, maintain that behind every sound school budget is a carefully structured and costed school development plan. It could be argued that a school's budget is not valid unless it is specifically tied to the school development plan – after all, how can a school claim to be planning to meet its stated priorities if it has not worked out how and when it will be able to afford them?

Development planning is a skill best acquired by the whole governing body together, along with the management team and key staff of the school. There is no way that it is realistic to expect that one individual can, unaided, write the whole school development plan and that plan be a legitimate statement of the school's aims and priorities. Yet there are schools where this, or something very similar to this, happens; and it is usually the headteacher, or a very small group of professionals, who produce the document.

But let's be reasonable about this. It is equally unrealistic to expect a governor, or even a group of governors, to produce the development plan. It is fair to expect that much of the administration work involved in formatting and producing the plan will be done in the school. However, the school development plan is the *governors'* strategic document. It is not the headteacher's or the deputy headteacher's personal management tool but the governors' strategic plan. It is therefore not only fair but essential that governors are fully involved in writing and reviewing the plan.

Good practice in some primary schools involves the use of some or all of a school training day.

Once a year in a particular school, all the teaching staff, governors, representatives of the non-teaching and administrative staff, and representatives of the parent body will come together to review progress on the previous year's plan and to set priorities for the next financial year. The headteacher or deputy will often start the day with a short presentation showing which targets from the previous year's plan have been achieved and will identify those targets that were not achieved. The presentation will also identify new initiatives that the school had not previously planned for, for example the introduction of the national Numeracy Hour.

All present will be split into groups representing a cross-section of interests and backgrounds within the school community. Each group will consider at least two sections of the plan, and their considerations will include new priorities alongside the outstanding priorities from the previous year. In addition, the groups may add other priorities if they think these may be important. Most sections of the plan will be reviewed and revised by at least two groups. Groups are asked to indicate levels of cost, a suitable timescale and a deadline for implementation, and to identify the person or group responsible for each target in that part of the plan.

At the end of the exercise, a brief plenary session gives each group the chance to explain its priorities. The notes from each group are then gathered and collated by the headteacher, deputy or (sometimes) a suitably knowledgeable governor. A small working group is convened to manage the development of these notes into a fully costed plan, ensuring that no individual or group is set unrealistic timescales for task completion or is responsible for too many tasks. In addition, the development priorities of subject areas, drawn up to a similar format by subject co-ordinators, are also added to the plan. The final plan, covering finance, buildings, curriculum, staffing, management and resources, is then considered by the appropriate subcommittees or working groups before being passed to the full governing body for approval.

In schools that adopt this or a similar approach, the governing body is well and truly in control of development. A governing body in this situation can be said to be following the four rules of successful development planning outlined in the box below.

The four rules of successful development planning

Priorities are to be agreed from a position of knowledge and involvement rather than simply rubber-stamping the priorities of one individual or one group of individuals.

All priorities identified are to make a difference to pupils' attainment.

Progress on all priorities can be measured so that the priorities are always clearly leading the school's target-setting strategies.

Each priority is to be costed so that governors can match expenditure against intended impact on pupils' attainment and, ultimately, measure the educational outcomes of their financial decisions.

The development of school policies is another area for which governors need careful oversight and full involvement because, like the school's annual development plan, the policies, once adopted, are the governors'. They are what the governors say the school stands for on areas as diverse as special educational needs and charging for school trips and visits. Governors need to know that, whoever drafts a policy initially, that policy goes through the appropriate working group or subcommittee for full consideration before being presented to the full governing body for approval.

Best practice often involves governors sitting with staff on small working groups that are responsible for producing particular policies. The policies are presented and discussed at staff meetings before going to the governing body (via the subcommittees). In some instances, governing bodies will send back policies on one or more occasions because they are not satisfied with a particular section, and these governors will never be caught out condoning action taken by the school against a policy that they have unwittingly, and without due consideration, agreed to. Where such vigilance has not occurred, governors have been known to get themselves tied up in hopeless knots because their own admissions criteria – not fully understood by governors themselves – have been circumvented, leading to real trouble at later appeals-panel hearings!

Curriculum

The governors are responsible for the curriculum that the school teaches its pupils. By law the governors have to have a curriculum statement and this has to include the arrangements for teaching religious education as well as the subjects of the National Curriculum. One could be forgiven for assuming that, since the advent of the National Curriculum and the subsequent years of fuss over how overcrowded the curriculum is, there is very little for the governors to decide in relation to the curriculum, given that most of it is statutory. There is some truth in this but the National Curriculum and religious education are not, and were never intended to be, the whole curriculum. In addition, governors have some scope in agreeing with headteachers what form the curriculum will take.

Here are four examples of flexibility in the curriculum:

- ❑ Drama never gained official recognition as a discrete subject in the National Curriculum but is intended to be taught mainly through 'Speaking and Listening' elements of the English curriculum. Some primary schools and many secondary schools still timetable drama lessons for their classes because it is felt that relying solely on the content of the English curriculum to teach the subject does not provide sufficient in-depth coverage or proper opportunities to develop pupils' skills and knowledge in drama to a higher level.
- ❑ At the time of the introduction of the National Curriculum, many secondary school governing bodies explored the idea of saving time on the curriculum through abolishing separate art, music and drama lessons and adopting a 'combined' or 'expressive' arts curriculum instead. Thankfully, the majority of schools decided against this strategy as it was generally unpopular with arts teachers and many parents because of the diluted experience that it would mean in art, music and drama for individual pupils, with a possible knock-on effect on examination results.
- ❑ Secondary schools timetable careers lessons and, in Key Stage 4, have the opportunity to offer a much wider range of courses than those mirroring the core and foundation subjects of the National Curriculum.
- ❑ Many schools timetable personal, social and health education lessons, although these are not mandatory under the National Curriculum.

A good school will keep its curriculum under constant review. The discretionary subject areas in particular will be monitored and their effect carefully measured. Statutory subject areas will be monitored to ensure that sufficient time is given over to them, that time-tabling arrangements are secure, and that the most appropriate frameworks are being used in which to teach the these subjects.

In addition, governors need to know that there are adequate and proper arrangements for managing the curriculum for those pupils with special educational needs. The Education Act 1996 and the *Code of Practice for Special Educational Needs* require children with special educational needs to have work prepared with their needs in mind. For those at Stage 1 (the lowest level of need), this may simply mean that the teacher adapts the class's work to make it accessible to these pupils (a practice known as 'differentiation'). Pupils on higher levels of need require Individual Education Plans that are drawn up specifically to ensure that the curriculum meets their needs.

Attainment and progress

The single most common measure of a school's success is the standards it attains for its pupils in national tests or public examinations. It is surprising, though, how many governors are not aware of their school's standing in English or mathematics, for example, against the local or national average. Best practice in schools means that governors are properly informed about these statistics (see Chapter 5). Every governing body should agree with the management team of the school a timetable for presentation of performance data, with an appropriate commentary, to the Curriculum Subcommittee or working group and then to the full governing body.

The presentation of attainment data is the easy part. Governors also have a need to know how well pupils are progressing. This involves more work for the school in analysing past data and in making valid comparisons, but it is essential that this work takes place if governors are going to be adequately informed for their target-setting responsibilities

Target setting is now statutory. Governors will have agreed literacy and numeracy targets for the year 2002 with their local authorities, which have in turn agreed their overall literacy and numeracy targets with the DfEE. Governors also have a responsibility to set annual targets for their schools. These will mainly be performance targets but do not necessarily have to be global targets. For example, raising the percentage of Level 5 attainment in mathematics at the end of Key Stage 2 may be a valid target if the school does well in terms of average attainment but shows little evidence of higher attainment. Other types of target may include reducing the overall percentage of absences or reducing the number of exclusions. These latter targets have implications for the whole school and the way that the school responds to issues such as absence and behaviour; but targets such as these will also impact on the attainment and progress of pupils who attend school more regularly and whose behaviour may be improved.

Quality-control strategies

Ask most headteachers what the single most important job is that a governing body has to undertake and the chances are they will say it is supporting the headteacher and the school. Support is best given from a position of knowledge and understanding of the work of the school. This knowledge and understanding requires the governors to have confidence in the quality-control strategies employed by the headteacher and the management team. Governors should not expect the headteacher to give them detailed reports on the capabilities of individual teachers but should feel comfortable that the headteacher has the capacity to identify teachers who may be struggling with parts of their work.

As the role of local education authorities changes to reflect a lower profile of local authority inspectors and advisers in more successful schools and a higher profile in less successful ones, governors of all schools need to know that there are good strategies in place for the regular and systematic monitoring of teaching and learning by management. Governors could expect, for example, to receive a report, perhaps through the Personnel Subcommittee, indicating how many lesson observations have taken place in the term so far, how many more are scheduled to take place that term, and whether all members of staff will have been seen by the end of the term. It may be useful to know the amount and nature of additional support that had been triggered by these visits.

Similarly, governors need to be confident that the work of pupils is being regularly and systematically monitored. It is usual for such monitoring to involve examination of a sample of work representing the average, below-average and above-average pupils of each class or year. A task as large as this is usually divided between headteacher, deputy headteacher and other senior teachers. In some schools, such sampling is carried out by subject co-ordinators and this sampling is monitored by management to ensure consistency. Whilst many schools have systems such as these in place, a knowledge of them by the governors is often scant. Some headteachers and their governors consider them professional issues that are not necessarily to be shared with governors; however, it is the content of the record of the observation or sampling – the detail of the visit – that is confidential, not the evidence of overall trends.

The system used to gather this information is very much the business of the governors. They, as the grouping ultimately responsible for the quality of education provided by the school, need to know that there are sound systems in place that can alert management to deficiencies in the quality of teaching and learning. They also have a right to know how management will respond to such issues when they arise. Without knowledge of these systems, governors are taking a big risk in trusting that their managers are doing the first-class job expected of them. With the knowledge that these systems are in place and effective comes the confidence to provide the support that the headteacher and management will need in times of difficulty.

Health and safety issues

Finally, governors need to know that their school is a safe and secure environment for pupils, teachers, other staff and visitors. The Education (School Premises) Regulations 1996 and the Health and Safety at Work Act 1974 place a requirement on governors to ensure the safety of pupils and staff. The Management of Health and Safety at Work Regulations 1992 require that risk assessments are regularly carried out. These cover every physical area of the building and every activity that the school is involved in.

The types of issues regularly identified are uneven playground surfaces, particularly where these are used for games lessons; wooden steps to mobile classrooms, damaged fencing or insecure gates leading to public rights of way, or obscure and insecure side and back entrances into schools. Many schools have undertaken a significant amount of work on external security since the Dunblane tragedy in 1996. Government inspection reports, being public reports, for obvious reasons do not include details of shortcomings in security, but these are raised by inspection teams directly with the headteacher and governors before the end of the inspection process. Whilst no school can ever be truly secure, selected governors should regularly 'walk the site' with staff responsible for health and safety with an eye to security as well as the material condition of the site.

In the classroom there are a wide range of potential hazards, often in the most unlikely places. Whereas one would expect a certain level of accidental injury in technology rooms, where materials such as wood, metal or plastic are being worked on and sharp tools are being used, the number of accidents in other contexts, such as art, often surprises people. Accidents often occur when pupils are cutting tiles for printing and are often down to pupils not observing safe practice in the use of tools. Batik equipment can also cause burns, and fixatives and some ceramic materials are toxic. Music rooms, too, can present the potential for injury as the number of electronic keyboards in schools increases without the provision of suitable furniture to house them. Often they are placed on tables at the start of each lesson, with mains cables trailing dangerously across the floor, and cleared away afterwards.

> One school had invested significantly in 10 state-of-the-art electronic keyboards. Within a year, all were out of use because the mains leads were damaged either at the keyboard end or at the mains plug end. Some leads had sustained damage through being carelessly rolled up after use, but others had been tripped on and pulled awkwardly out of their sockets, putting the safety and well-being of children at risk as well as causing damage to valuable resources.

The foregoing are just a few examples of the type of health and safety issues that can arise in school, but every building and every curriculum will present its own set of health-and-safety challenges. Governors are required to have a policy on health and safety and informed governing bodies will, through their Premises or Resources Subcommittee, receive outcomes of risk assessments carried out and will be informed of any urgent issues that need addressing.

Summary

Table 6.1 summarizes the main points from each of the above paragraph headings. A good governing body will know that the knowledge of the school, as identified in column one of the table, is shared amongst the governing body as a whole and will know which governors have a lead on information and decision making in each area. The second column of Table 6.1 should help governors identify:

- ❑ where responsibility for a particular issue falls;
- ❑ where questions on specific issues should be directed by other governors;
- ❑ who should produce reports on these areas.

Table 6.1 *A summary of governors' responsibilities*

WHAT GOVERNORS NEED TO KNOW	WHICH GOVERNORS NEED TO KNOW FIRST
Personnel matters, staffing structure, job descriptions, appraisal, professional development	Personnel Subcommittee
Finance, school budget	Finance Subcommittee
Accommodation and resources	Resources/Premises and Finance Subcommittees
Pupil information (socio-economic circumstances, special needs, etc)	Personnel Subcommittee
School development plan	All
Policy development	Finance – all policy development issues. Other committees as appropriate
Curriculum issues	Curriculum Subcommittee
Attainment and progress	Curriculum Subcommittee
Target setting (standards)	Curriculum Subcommittee
Target setting (planning)	All
Quality control (teaching/learning)	Personnel/Curriculum Subcommittees
Health and safety	Finance/Resources and Premises Subcommittees

WHAT HEADTEACHERS NEED TO KNOW

Most headteachers would be expected to know everything that was included in the previous section for school governors. In larger schools, the management team would hold the knowledge collectively. However, whereas there are many areas of overlap there are also different levels of depth of knowledge required by managers, and some areas of knowledge will be approached from different perspectives by managers than by governors. Whatever other information a headteacher or management team may hold about their school, this section identifies that for which good systems are required for keeping valuable information up-to-date. Once again, this is essential if school managers are to work effectively with their governors and local education authority in setting appropriate targets.

How well teachers are teaching and planning

The importance of effective systems for monitoring the quality of the work of teachers cannot be underestimated. Teachers are the ingredient that makes a school effective, and good management is the ingredient that makes teachers effective. Management can only be effective if it has information on the strengths and weaknesses of its teaching force and if it can then use its resources to build on the strengths and address the weaknesses. The larger a school gets, the more important it is to have a structured system that several managers will contribute to but that will be overseen by one manager. The smaller a school gets, the easier it is for a rigorous system to become compromised because the one or two individuals undertaking the monitoring may allow the time for monitoring to be squeezed out by other 'priorities'. There is a Classroom Monitoring Checklist shown in Figure 4.2 (Chapter 4), but Figure 6.1 should assist schools in setting up and keeping a record of regular visits to individual teachers.

Records of observations of teachers' work are confidential and should not be shared with any other member of staff, any governors or any other party without the consent of the teacher involved. In the best systems, feedback to the teacher will take place later on the day of the observation and at the feedback the observing manager would share written notes with the teacher. A record of the observation and agreed targets to be revisited at the next scheduled observation will be kept on file and a copy offered to the teacher. In addition to the regularly scheduled visits, there may be a need for additional visits if particular difficulties requiring closer monitoring are identified – for example, a teacher experiencing poor class control. These extra visits should be recorded in the same way, together with a record of the additional support that the management team is arranging in order to help the teacher overcome the difficulties. It is true that such records can help form a relevant evidence base in the event of disciplinary proceedings being taken out against a teacher, but the records will also need to show the level of support given to the teacher after problems have been identified and the commitment of management to help the teacher overcome these difficulties. The primary purpose of such a system is to support and develop and not to gather evidence for possible future disciplinary action.

Monitoring Activity Sheet					
Teacher's name:					
Main responsibilities:					
Date of observation	**Lesson/class or context observed**	**Carried out by**	**Notes shared on**	**Targets agreed on**	**Follow up visit on**
Autumn term					
Spring term					
Summer term					
Notes					

Figure 6.1 *A sample Monitoring Activity Sheet*

A systematic strategy for monitoring the work of teachers in classrooms gives managers the confidence that they need in their teaching staff and guards against unpleasant surprises in the future. There are headteachers who go into inspections by the national inspections agency seemingly unaware of poor teaching in their schools. The majority of criticism in these instances in not levelled at the teacher but at the managers who have not picked up the problem and supported the teacher.

The monitoring of teachers' planning is an essential activity that should take place throughout the school year, and Figure 4.1 in Chapter 4 shows a Planning Monitoring Sheet used within a comprehensive system that covers the monitoring of long-term, medium-term and short-term planning. It is suggested that a copy of this record is kept at the beginning of each teacher's planner so that the monitoring system is transparent to the teacher.

Whether teachers' assessment records are up-to-date and useful

One of the most inconsistent practices seen in many schools is the recording of pupils' progress and attainment. The introduction of national testing has done much to improve the situation by providing teachers with a clear indication of what must be measured and by expecting a minimum amount of measuring to take place for teacher assessment. However, this system only applies to the core subjects of English, mathematics and science.

The National Curriculum requires reporting at the end of Key Stages to be made to parents against specific criteria. This implies that schools should keep records of pupils' progress against these criteria for all subjects, so as to enable the required reporting to take place. Realistically, there is a limit to the amount of time that teachers have available for testing pupils and for ticking boxes. This means that the systems devised have to be simple, be non-text-based as far as is possible, and provide the maximum information for the minimum effort.

Many schools have devised generic schemes, into which the elements of the attainment targets of each foundation subject can be inserted before copying the form for completion for each child. Other schools are experimenting with computer-based systems, which offer an enormous power and flexibility in interpreting and presenting the information. Any system that generates the main content of an end-of-term or end-of-year report automatically for each child from assessment data easily entered during the course of each term has to be worth serious consideration.

If schools do not have whole-school systems for recording pupils' progress, it is quite difficult to know whether or not accurate and relevant records are being kept. Many primary schools still rely on teachers keeping their own informal records to be used later for assessing progress in the core subjects. There are several dangers with this practice: firstly, that the quality of each teacher's records cannot be guaranteed – teachers may be interpreting broad criteria differently; secondly, that the consistency with which teachers update these records cannot be guaranteed, for it is tempting to take short cuts on informal systems from time to time if there are other priorities; finally, if teachers leave mid-term or mid-year, the records may not be up-to-date or may not be easily understood by the replacement teachers.

The accuracy of teachers' assessments is an area that good schools pay significant attention to as part of the assessment process. When working in isolation, teachers often find it difficult confidently to 'level' (ie grade) a piece of work. In good schools, agreement trialling or shared exercises will take place periodically to ensure the consistency and accuracy of assessment; in these schools the accuracy and effectiveness of this system is also checked by management, through sampling assessments, as part of their monitoring process.

What are the training needs of staff?

The monitoring of teachers' work and the examination of planning can reveal whether or not training or support is needed on classroom organization or on developing the teachers' knowledge or expertise in a certain area. In addition, as part of regular professional interviews, teachers should have the opportunity to identify areas in which they are interested in developing further through training.

Good practice in the management of teachers' professional development would mean that a school takes individual needs into account alongside institutional needs. For example, there may be a need to train all the staff of a primary school in implementing the Numeracy Hour, but there may also be a need to provide additional training for one or more teachers in the area of Using and Applying Mathematics (Attainment Target 1).

Good practice will also mean that a school has a system for assessing the impact of addressing a teacher's own identified needs against the overall target of improving attainment in the school. It may well be that a teacher who wishes to attend a course to develop library skills may be a lower priority than the teacher who needs support in the area of Using and Applying Mathematics, because management considers that the impact of providing support for the latter will be greater on the overall attainment level of pupils than the impact of providing support for the former.

A system that helps to weight the training needs of each member of staff against the school's overall aims and objectives will be more effective in improving the overall performance of the school than a system that shares professional development opportunities out equally between all staff. Any system should reflect the priorities outlined in the school development plan but should also have a contingency for providing support beyond these agreed priorities. Of course, staff have a right to further training and care needs to be taken that good and high-performing teachers are not deprived of opportunities to keep their skills up-to-date, to develop them further, or to acquire new skills.

How adequately is the school resourced?

Good schools have developed the role of curriculum co-ordinators and heads of department sufficiently to ensure that these individuals have a clear picture of the level of resourcing in the areas for which they are responsible. Many schools have reasonable systems in place for monitoring expenditure in curriculum areas, particularly monitoring the usage of consumable resources and, to a lesser extent, books. However, the approach to the development of other long-term or permanent resources can often be less secure.

Different subject areas have widely differing needs in terms of resourcing of equipment. As schools begin to broaden the use of information technology across the curriculum, those needs will become even more diverse as the uniformity of hardware is contrasted with a diversity of software and peripheral devices in some curriculum areas. Managers have a far better chance of being able to keep subject areas adequately resourced if they have access to a central and up-to-date database of key resources across the school, together with an indication of the following for each department:

- ❑ sufficiency of resources;
- ❑ quality of resources;
- ❑ maintenance costs;
- ❑ likely replacement date;
- ❑ new equipment needed or desired within subject areas.

In addition to knowing the level of required resourcing and being able to plan to sustain and develop this, schools need to know how well resources, such as the library or other information resources facility, are used. This information is becoming increasingly valuable to schools in the target-setting process, where it may as a result be possible to correlate weaknesses in aspects of reading or poor research and retrieval skills with low levels of library usage. Similarly, there is a risk of significant underuse of high-quality IT resources in some schools because teachers in other subject areas, or without skills or knowledge in the area of information technology, may not know the potential of such equipment in supporting their work.

There is one fundamental question that should be asked of all teachers and managers before they spend anything on additional or new resources. That question is simply: 'What difference will this resource or expenditure make to pupils' attainment?' If the answer to this question does not come easily, then questions should be asked as to whether or not this expenditure is justified. We all know that literacy, for example, is a national priority; we also know that information and communications technology is another national priority. However, teachers and governors understand the processes involved in teaching and learning in literacy far more than they do in information and communications technology, and so schools find it easier, given the availability of resources, to spend money on improving library stock and facilities. Have they, though, considered the effect of this expenditure on further raising standards in literacy? What if standards are already high: are the schools convinced that standards will be raised further? Did they consider that they may have had a greater need in investing in training and resourcing in information and communications technology? Did they even discuss the issue?

By reading many national inspection reports, it can be seen that expenditure on literacy in schools where standards are already at or above the national expectation will almost always outstrip expenditure on information and communications technology where standards are below the national expectation. Provocative? Yes – but as we move into the third millennium AD should we not be looking hard at the way we resource our schools and the way that we teach our children to learn? Hand in hand with the teaching of literacy should be the equally rigorous, equally well-resourced and equally imaginative teaching of information and communications technology. Adults without good skills in reading, speaking and listening will find it difficult to progress in life, but this rule applies equally nowadays to those who cannot access and use technology to help themselves, their families and their businesses.

How effective are the arrangements for pupils' pastoral care, social development and child protection?

Pupils will always be better learners in an environment that is secure and positive.

> An inspector, who had been a pupil of a boys' grammar school in the north of England, remembers vividly that bullying there was not just rife, it was a way of life. It wasn't only the bullying from other boys that made life difficult: at best, teachers knew it went on and did little to tackle it other than to encourage the bullied to stand up for themselves; at worst, some teachers were bullies as well using control mechanisms that were highly suspect. We are not simply talking about corporal punishment here but strategies such as electric shocks in the physics laboratory or random punishments based on the Roman disciplinary regime of decimation (where every 10th pupil might arbitrarily receive a punishment). The worse thing was that the boys accepted unjust punishments as part and parcel of growing up and being at school; they assumed that this was how it was in any school.

> On inspecting a boys' school recently, the inspector found that the macho environment that he had expected to see was not evident. That isn't to say that bullying didn't exist: it did, but on a far smaller scale than he had expected and as only a small distraction in an environment that was positive and caring overall. This school undoubtedly had developed a value system that allowed boys to respect each other's strengths and weaknesses without being called 'soft'.

Pastoral care

It is recognized nowadays that pupils bring to school a range of circumstances and backgrounds that can affect their learning. Schools have support mechanisms to address many of these instances and there is a range of support services available to schools to help with more acute difficulties. There is nonetheless some risk that schools will begin to blame all poor attitudes to learning or antisocial behaviour on the pupils, their backgrounds or their home circumstances.

First and foremost schools must remember that, whilst there always will be children who need support and help, the school's own weak teachers, including those who may be good lecturers with high levels of subject knowledge but poor communicators with children, may well be at the root of many pupils' reluctance to learn or comply with the school's behaviour code. As society itself becomes more sophisticated, children too are becoming more selective and are less likely to accept poor-quality teaching in return for their presence and undivided attention at school. We live in a choice-driven society, and education is far from immune from the developing trends for increased consumer choice. Parents are constantly being urged to choose the better schools for their children. Should we be surprised, therefore, when some of these children choose to spend their time in what they see as more constructive pastimes or activities rather than attend poor-quality lessons?

A headteacher needs to know that the strategies for pastoral care are not a substitute for poor teaching elsewhere in the system. The development of systems that show how effective pastoral care improves attitudes and attainment is vital in developing this knowledge. Once again, personal target setting, linked to academic performance as well as to behaviour or attendance, has a part to play in this process. Schools need also to undertake an analysis of support needs and consider other school data, such as that on teacher effectiveness, as a part of this process.

Personal and social development

One of the 'poor relations' of the non-statutory curriculum is personal and social development. Often, this subject area is headed up by an enthusiastic and appropriately trained individual but is then let down by teaching staff who are apathetic (towards this area of the curriculum, at least) and who have little understanding of, or interest in, yet another subject that they have to teach. In primary schools, teachers have the advantage of being with their class for almost the whole week and an imaginative scheme for personal and social development can, with good training and appropriate resourcing, be easily integrated into the school day. Secondary schools may rely on form tutors to teach the subject, often with minimal training. For teachers who came into the profession to teach mathematics or art this can be a problem. It is less of a problem for teachers who teach children mathematics or art and who recognize that good personal and social development can contribute to learning.

Other secondary schools may teach personal and social development through subjects such as science, religious education, and drama. In these schools a good co-ordinator will work with the heads of departments to ensure that the subject schemes of work address issues of personal and social education adequately. One advantage of this system is that personal and social education is not being taught in isolation. Because the aim of the lesson can be reached through specialist subject teachers teaching their own subjects well, the chances of a more consistent quality of provision are perhaps higher.

A good manager in a school will consider seriously whether or not the school gives sufficient support to the personal and social development of its pupils and whether this area has sufficient priority in the eyes of the management team and the teachers.

Child protection

There are statutory requirements for schools to have child protection policies and procedures. The effectiveness of these, because of the nature of the area, can be difficult to evaluate. However, a school can and should regularly review its systems to ensure that procedures are clearly documented and known by all staff, that staff are appropriately trained to recognize signs of possible abuse and to know how to respond to pupils who may approach them with such difficulties, and that pupils know that help and advice are available within the school.

How effective are the arrangements for behaviour management?

Schools are complex organizations, and weaknesses in the systems to support one area of the school's work may threaten the successes being achieved in other areas. The relationship between pastoral care provision and behaviour management is an obvious one. If both systems are working well, then the chances are that the time spent by teachers in implementing each system will be less than if one or both systems were not working well. For example, if pupils are being bullied and poor behaviour management by some teachers is not dealing with the problem, this will create additional pressure on the pastoral support system. Equally, if the school does not teach and promote respect for the feelings, beliefs and property of others, then it is more likely to have behaviour-related problems. Such interactions impact on the learning of pupils in the system – sometimes significantly so on a large number of pupils. If learning is thereby being held back, standards of attainment will never be improved.

Most schools have developed codes of practice for pupil behaviour. Visits to classrooms in different schools can identify, displayed on the wall and sometimes in pupils' books, Golden Rules, Class Agreements or, in some cases, the teachers' or school's expectations written out in a rather direct and no-nonsense way. It doesn't matter what is displayed on the walls, though; what matters is that there is a system that is:

- ❑ seen to be fair;
- ❑ known and understood by all pupils;
- ❑ known and understood by all staff; and, most importantly,
- ❑ consistently applied by staff.

Managers need to know that all staff are fully trained in the school's system for challenging poor behaviour. They need to know that the school's policy on sanctions and rewards is consistently applied. This means keeping records that allow the use of rewards and sanctions to be properly monitored. It may mean talking to samples of pupils who have been the subject of rewards or sanctions about their perception of the fairness of the system. It will mean challenging staff who are not using the system or who are abusing it. It also means ensuring that the systems are regularly reviewed and updated in light of their effectiveness as judged through the monitoring exercises.

How effective are the arrangements for addressing poor attendance and promoting punctuality?

One of the first requirements for successful learning is to get the pupils into school every day on time. There has been much talk about the need to expand the school day, but many schools have discovered that if the time available is used more effectively in the first place then there is no need to extend the school day. The following examples illustrate how, in weaker schools, simply tightening up on existing systems can produce more time for learning.

One primary school had parents turning up with their children up to 40 minutes after the official start of the school day. The same school had parents withdrawing their children during term time for off-peak family holidays. One parent, when challenged by the new headteacher on her child's persistent lateness, said 'This is why this school is such a lovely school – nobody bothers what time the children arrive!'

The new headteacher has since made significant progress in educating parents to take a more responsible attitude towards their children's schooling. Lateness has significantly declined and term-time holidays are more infrequent.

Another school found that it could increase the amount of time spent in teaching and learning activities if it implemented the starting times of sessions and lessons more strictly. Staff were challenged if they were late arriving at their classrooms, and pupils were challenged if they were not moving purposefully in corridors between lessons.

The result was that 40-minute lessons, thought to be far too short by many staff, were seen to be quite adequate if the time was used effectively. This was particularly so when teachers were well organized and prepared for their lessons and were able to commence effective teaching straight away.

New government initiatives are helping schools to address attendance issues through target setting. In good schools this target setting is supported by careful analysis of reasons for poor attendance, co-operation with education welfare officers in addressing the issues, and strategies to support pupils who have poor records of attendance or who are at risk of becoming poor attenders. Regular reviews of the quality of these contacts and of the effectiveness of school strategies to improve attendance are important if improvements are to be made. Similarly, schools need to monitor timekeeping during the school day to ensure that both teachers and pupils are making best use of the available time by being where they are supposed to be – on time at all times.

How effective is the provision for special educational needs?

The *Code of Practice for Special Educational Needs* gives all schools a proven and effective framework for the identification and management of pupils with special educational needs. It also ensures that all schools have a special needs co-ordinator and a named governor with a specific responsibility for overseeing special needs. Because of this system, it can be easy for managers to leave the special needs co-ordinator to get on with the job with minimal interference. Whereas there are merits in this approach with any effective staff member, the management still has a responsibility to ensure that the systems used by the school are identifying the correct pupils and that these pupils are receiving the correct support.

Regular review meetings, required by the Code of Practice, should ensure that pupils are assessed at the right level. Managers should nevertheless take a particular interest in the Individual Education Plans drawn up for each child at the appropriate level on the special needs register. It is these plans that encapsulate the school's response to a child's specific need and that describe how the need will be addressed. If the plan is not right, it will not work; if the plan lacks appropriate detail, it will not be as effective as it could have been; if the plan is not implemented properly by the pupil's teachers, it will serve little purpose. There needs to be rigorous monitoring of the formulation and implementation of individual education plans regularly throughout the school year.

Managers also need to know that the time allocated to their special needs co-ordinators is realistic in recognizing the demands of the job. Pupils with special needs deserve an appropriate share of the school's resourcing if their own contribution to the overall picture of rising standards in the school is to be recognized.

Summary

The purpose of this section is not to provide an exhaustive list of necessary knowledge that should be in the possession of all headteachers. It is to identify the key areas within a school that must be properly monitored and supported if management is to be seen to be being effective. Table 6.2 illustrates these areas and the type of strategy that can be used to inform the headteacher as to their status.

Table 6.2 *A summary of headteachers' responsibilities*

WHAT HEADTEACHERS NEED TO KNOW	WHICH SYSTEMS CAN HELP
How well teachers are teaching	Direct monitoring through observation
How well teachers are planning	Monitoring through regular and systematic examination of plans
Is pupil assessment up-to-date and useful?	Monitoring through regular and systematic examination records; through sampling pupils' work and through considering assessment in teachers' planning
What are the training needs of staff?	Identified system to balance institutional and personal needs, including appraisal
How adequately is the school resourced?	Identified system to evaluate resource need, likely effect of resourcing on attainment, and value for money
How effective are the arrangements for pupils' pastoral care, social development and child protection?	Regular review and monitoring of arrangements. Good induction procedures and targeted training
How effective are the arrangements for behaviour management?	Monitoring of incidences of poor behaviour and use of school's rewards and sanctions systems
How effective are the arrangements for addressing poor attendance and promoting punctuality?	Monitoring of incidences of lateness or non-attendance and the effectiveness of school's strategies in these areas
How effective is the provision for special educational needs?	Monitoring of implementation of the *Code of Practice* and the quality and use of individual action plans

WHAT TEACHERS NEED TO KNOW

Essential knowledge for teachers to help them do their job effectively can be divided into two areas. The first concerns the school's systems – systems that are there to support them and ensure a consistent approach by all staff to the needs of their pupils. The second concerns the pupils and includes what the teachers teach, how they prepare and deliver their lessons, and how they record and report progress and attainment.

This section assumes that teachers follow the systems that the school has adopted for pastoral care, planning, assessment, marking, rewarding good conduct and effort, and addressing poor performance. It also assumes that teachers have appropriate subject knowledge and knowledge of how pupils learn. However, if teachers are to be effective, they need to know that they are using the right strategies in their planning and teaching.

How effective is the teaching and planning?

When the system of regular national school inspections was introduced, it came as a real shock to many teachers. Nobody particularly likes having their work directly observed, and for many teachers the experience of having an individual in the classroom, making judgements about their performance, was totally new and very unwelcome.

> The author remembers his experience of being visited by Her Majesty's Inspectors of Schools some years ago. The observation itself was not so bad – a particular project being run by the school was attracting significant attention, and so visitors to the classroom were not unusual. What was difficult to handle was that the inspectors arrived without saying anything, didn't communicate during their visit, and then left without saying anything. Worse still, there was no feedback – none at all.

Monitoring teaching

Whatever the shortcomings are of the new system of school inspections, teachers should be able to establish some sort of contact with their inspectors if they wish, and all teachers are entitled to some feedback.

Teachers now have the right to complain if they believe that inspectors are not treating them fairly – and they should exercise that right if need be.

Many schools have also now implemented some form of internal monitoring. Young teachers particularly are used to having their lessons observed, and many have told me that they welcome a professional opinion of their work. Headteachers used to find monitoring teaching quite difficult, particularly managing the feedback; some still do. Yet it is the feedback that is the most important part of the monitoring exercise. There is absolutely no point in gathering information on a teacher's performance unless that information is going to be used to improve that teacher's performance. If it is not shared, improvements can rarely be made.

Feedback should be prepared carefully. A good set of procedures for monitoring the lesson will make this easier. In some schools, feedback on lesson observations is combined with a wider professional discussion on other performance indicators. As long as there is no significant delay between the lesson observation and the feedback, this is acceptable. In ideal circumstances the lesson feedback should take place as soon as possible on the day of the observation. If the feedback is to be a part of a wider interview at a later date, a summary should be given to the teacher on the day of the observation.

Good feedback will include the following elements:

- ❑ the teacher's own perception of the success or otherwise of the observed lessons;
- ❑ the main strengths of the teaching as noted by the observer;
- ❑ the main weaknesses, or areas for development as agreed by the observer and teacher;
- ❑ a clear explanation of why weaknesses are thought to be present, if there is no agreement on particular points;
- ❑ identification of specific targets for improvement within a given timescale;
- ❑ a commitment to providing identified training or support if necessary in order to help meet the targets set;
- ❑ a summary of the feedback, finishing with a positive message and with mutual satisfaction that the experience has been productive.

Monitoring planning

If teachers' planning is regularly monitored, as it should be, then it is essential that teachers are given appropriate feedback on its effectiveness. This can and should happen whenever lessons are observed.

However, these observations generally only take place termly at most for the majority of teachers. Good planning does not guarantee good teaching, but it is a significant help in supporting good teaching. Schools where planning is monitored on a weekly basis know this, and schools that use a monitoring system, such as the one described in Chapter 4, are supporting their teachers through showing them what is being looked for in the monitoring exercise and identifying weaknesses if they are found.

What have pupils learned and what exactly do they need to learn?

Teaching is at its most effective when it builds on what pupils have previously learned and experienced. In order to do this successfully, teachers have to have accurate records of each pupil's attainment as well as a knowledge of their weaknesses, strengths and specific factors that may influence their learning. Record keeping has always been a problem for teachers. Done properly it requires a file for each pupil similar to a GP's medical file on each patient. However, the profession learned in the early 1990s that it was not possible to assess and record each pupil's progress in every subject of the curriculum, even though there is still a vague expectation expressed in national inspection reports that they should do so.

Undoubtedly technology holds the answer, and within a few years teachers will be able to enter into a terminal on their desk single keystrokes that will inform a central database of every aspects of a child's learning and progress. Until that time, we have to make do with paper-based systems that rarely leave an individual teacher's classroom and that are not sufficiently shared with colleagues. It is unfortunate that the very technology that could ease the burden of insufficient time suffered by many teachers is not becoming accessible to the profession quickly enough because of a combination of short-sighted management and lack of time to learn to use the new technology.

One of the easiest ways accurately to record learning for individual pupils is to combine paper-based record systems, of the type that allows a teacher to enter a mark to show the status of a child's learning against a strand of a subject of the curriculum, with good planning. A well planned lesson, set of lessons or module, will show clear learning objectives not only for the whole class but for different ability groups within that class. Each learning objective is an assessment criterion in itself and can be used to record a child's success in meeting that criterion. If a learning objective is not met, it should be addressed in a different way after the teacher has identified why it has not been met. (How many teachers actually look at learning objectives they have written for a lesson *after* they have taught the lesson and ask whether they have been met?)

Good planning practice will identify key learning objectives in the medium term. Specific learning objectives should be identified for different ability groups in the shorter term. Evaluating and recording the success of learning against these objectives for each group helps confirm what pupils have learned and helps provide input to the planning of the next lesson or module. Used with an assessment system that incorporates the planned learning objectives and occasional notes on individual children, the system is as comprehensive as it needs to be. Of course, evidence needs to be kept, in the form of samples of pupils' work, and annotations to this evidence can provide further information.

Are the pupils making progress?

This is the single most important question that a teacher needs to ask – every day and in every lesson. It is not sufficient to have good systems for recording what pupils have learnt and to use those systems to ensure that learning is progressive. It is essential that the records show that progress is at a level that shows that pupils are being properly challenged.

Schools are beginning to learn pupil-tracking procedures that allow them to follow a child's progress from the end of one key stage to the end of another. There are recognized and well-publicized indicators to confirm whether, over time, progress has been satisfactory. Use of these systems will allow schools to identify potential weaknesses in the years between the two sets of assessments and try to address them. The problem with this method is that it is too late for the children whose performance may have been identified as being weak; something needs to be done over a shorter timescale – within the school year, for instance – to give teachers confidence in knowing that their pupils are progressing at an appropriate pace.

Targets are a useful answer to this problem. Good planning will clearly identify, over the longer term, the knowledge, concepts and skills needed successfully to make progress in a particular subject – progress that can be measured through National Curriculum level descriptors, for example. Medium-term and short-term planning will identify the learning objectives needed to ensure that the appropriate knowledge, concepts and skills are taught to all pupils. Another word for learning objectives is targets, and general learning objectives are targets for the whole class. Short-term learning objectives may be targets for specific groups of pupils. Failed learning objectives become targets for individual pupils and could be addressed as such: if a child does not achieve a particular learning objective, then that child needs support in helping him or her achieve the objective over time.

This technique ensures that planning, teaching and assessment are not carried out in isolation. It gives a clear purpose to each aspect and helps teachers identify quickly the planning tasks that are relevant and those that are there through habit – the latter can be dropped. Concentrate on making the learning objectives happen and anything else in the planning can be an unnecessary extra.

Ensuring progress is a complex activity. There is little point in having effective and detailed planning strategies if the learning objectives are missing or weak. The planning will give a teacher peace of mind in the lesson but that teacher will never be really sure of the actual outcome in terms of effective learning because the real aims of the teaching were never identified in the first place. The result could be a well-organized and well-presented experience but an experience of unknown value! Similarly, there is little point in having clear learning objectives in planning and teaching if there is not going to be a record of whether or not the pupils are achieving those objectives. The three strategies are essential and interdependent. Rather like the old three-legged stool, take one support away and the whole process is in imminent danger of collapse.

Summary

Teachers need information on their performance to enable them to improve. Rather like the newly promoted manager who asked his line manager how he would know if he was doing a good job, teachers need more than the response 'We'll let you know if you mess up!' Everybody needs encouragement in their work, and the greater majority of employees – wherever they may work – welcome constructive advice, particularly if it will make them improve their performance and perhaps enhance their career prospects. Similarly, teachers who know and can show that all of their pupils, regardless of their abilities, are making progress will feel much more confident in the effectiveness of their work and will serve their pupils much better. Table 6.3 gives the basic framework for keeping teachers informed and about their own performance and that of their pupils.

Table 6.3 *A summary of teachers' responsibilities in the classroom*

WHAT TEACHERS NEED TO KNOW	WHICH SYSTEMS CAN HELP
How effective is the teaching?	Regular monitoring by direct observation of lessons, at least once a term, by a line manager and by the headteacher. Early and sensitive feedback with strengths and areas for improvement clearly identified. Targets for improvement to be reviewed within a given timescale
How effective is the planning?	Weekly systematic monitoring of their planning by a line manager using known criteria. Feedback on the quality of planning and areas for improvement
What have pupils learned and what exactly do they need to learn?	Recording systems that are low-maintenance yet information-rich. Assessment systems that are linked closely to identified Learning Outcomes in teachers' planning
Are the pupils making progress?	Planning and assessment systems that are complementary and allow easy tracking of pupils' progress. Target setting and review systems

WHAT PARENTS NEED TO KNOW

It can be very difficult being a parent with an interest in your child's education. Despite the rhetoric about home–school partnerships and high levels of accessibility to headteachers and teachers, the experience of many parents is still that it is quite difficult to get access to a school when they want it.

Establishing contact

In Chapter 1 a description was given of a school with a particularly thoughtful and imaginative way of enabling parental contact with the headteacher and staff. The system works for that school because the headteacher believes that problems are best dealt with quickly and early. There is no greater time commitment given over to meeting parents using this method; it is simply that the time is given when the parent wants it and not when the school thinks it should provide it.

Examination of the customer relations statements and policies of many major companies shows that they value the customer's desire to be responded to quickly and courteously. Firms as diverse as hi-fi chains to exhaust and tyre companies display notices reminding staff and customers that the customers are the most important people in the shop. If prompt attention is seen as a necessity when your CD player breaks down, then surely the same prompt attention should be given to you when there is a problem with your child. Even visitors to the DfEE are given evaluation forms to complete in order to indicate the level of service they have received from the department. Few schools seek this type of feedback from its parents on a systematic basis.

All too often a parent or guardian does not have telephone calls returned, is told that the individual being sought is unavailable and is not offered an alternative, or is given an appointment several days into the future. Parents usually only contact schools when there is a real problem. Often the point that the parent recognizes the problem is a point well beyond that when there should have been communication between home and school. To be offered an interview in several days' time in these circumstances is seen by the parent, quite justifiably, as a wholly inadequate response.

Many parents are aware that the first time they contact their child's school they are exposing themselves as minor irritations. If they contact the school again, they are becoming nuisances. God forbid that they contact the school a third time and become the 'parents from hell' who are the talk of the school office and staff room. It's a funny situation when caring parents can be seen as bigger nuisances than the bullies who are making their child's life a misery!

Whatever system schools choose to use to manage parental contact, they must at least try to make it look as if they value that contact and welcome the opportunity to talk to parents. The provision of contact numbers, names and, if necessary, times when it is appropriate to contact staff should be shared regularly – perhaps even appended to reports or curriculum information. Furthermore, even before schools set up such systems as that described in Chapter 1, there is a lot that can be done to ensure that parents have sufficient information both to keep the need for personal contact to a minimum and to ensure that parents know how they can contact the school if there is a necessity to do so.

Being aware of a child's progress

Schools have a statutory obligation to send to parents reports on their children's attainment. However, although this information can be useful, it rarely tells parents with a real interest in their child's learning anything at all about their progress. The fact that 'Jonathan can read books appropriate for his age with fluency and expression' is probably not news for his mum and dad – Jonathan has probably been demonstrating his excellent reading skills for several months! What would be really useful to his parents is a comment on how much his reading has improved since the start of the year or the time of the last report; and not just a comment but a statement describing the evidence to support the teacher's view that Jonathan's reading skills are improving. This would then be something that reassures his parents that the school recognizes that their child learns quickly and needs challenging in his reading.

Reports to parents should not simply state the obvious. They should provide a context for future developments – developments that can be explored at parents' consultation meetings, when the parents role in moving their child forward can also be discussed.

> The parents of one seven-year-old were told by his school that he had a gift for art. Naïvely the parents waited eagerly for the next report, which they were sure would show that the gift was being nurtured. Neither were artists themselves, and so they found it difficult to tell whether or not his drawing was improving. In addition, the child did not seem particularly interested in the subject when art-related gifts were bought for him.
>
> The teacher had recognized talent but did not have the expertise herself to move it forward. The next teacher neither had the expertise or the ability to even recognize the talent. The school had no art specialist who could help. Art was never mentioned on the child's report again at primary school.
>
> It was only several years later at secondary school when the talent was picked up again, but who knows what opportunities had by then been lost for this individual. Better communication between the school and parents could have ensured that the child's progress in this subject was carefully monitored and that expert help was brought in if this was necessary.

Schools are developing banks of information that give them indicators about their pupils' progress. It is reasonable to expect that this information, simplified and placed in context where necessary, should be shared with parents. Not only national assessment results but entry profiles, reading-test scores and other forms of regular testing that, with reference to tests previously carried out, can be used to report progress.

Being aware of teachers' concerns

Annual reports and termly or twice-yearly meetings with parents may be sufficient to ensure the parents of most children that reasonable or good progress is being made and that any minor problems have been recognized and are being addressed. However, it does come as a shock to many parents to be told at a parents' consultation evening mid-way through the year that their child is seriously underachieving.

A reasonable response from the parent would be 'Why on earth didn't you tell me earlier?' Another reasonable response would be 'What are you doing about it then?' It is not enough for a school to tell a parent that their child is not working as well as that child could be doing. The school needs to win over the parents' confidence and co-operation by explaining a strategy for getting the child's performance back to where it should be. In addition, the school should really be initiating the kind of meeting where this type of information is being shared at a much earlier stage. It doesn't take long for bad habits to set in with any child, but it takes a lot longer for those habits to be corrected, particularly for a child in Year 9. To be told at Christmas, or later, that your child has wasted an entire term is simply not on. A third of a whole school year is significantly more time than any child can afford to lose out of education.

One secondary school that was asked by a child's parents 'What are you doing about it then?' talked half-heartedly about parents' responsibilities. The parents came straight back and said that they checked and signed their child's homework diary every night and looked at the homework regularly.

They asked again, 'What is the school doing about it, because we are being supportive?' In this instance, the school's response was magnificent. A non-punitive programme of support and targets was set for the child, whose attitude and progress improved significantly throughout the remainder of the year. The parents doubted that this would have happened had they not challenged the school.

Schools often complain that they do not get the support of parents. Sharing concerns at an early stage and sharing strategies to address these concerns is one way of ensuring parental support. The annual report and meeting should not be the vehicles for exposing serious concerns about significant shortcomings in a child's attitude, progress or behaviour; they should confirm what the school and the parents *already* know about the child.

Knowing what teachers are going to teach each term

Sharing curriculum information with parents is a task that all schools are required to do, but few do it successfully. In fairness to schools, it is not always their fault. Schools can spend hours on preparing science evenings or mathematics evenings, only to have a small handful of interested parents turn up. The school prospectus is the document that has to contain curriculum information. However, the information is often very bland, rather obvious, full of self-praise and gratuitous description, and only read by prospective parents and not by the parents of those pupils already at the school.

Primary schools are often much better at sharing information than their secondary counterparts. Some schools publish a termly sheet for each class, which covers the main areas of each subject being taught. Many, however, limit the publication to classroom windows or parents' noticeboards. This strategy is of little use to parents of older children or younger ones who are brought to school by friends for example. Good practice is to produce the concise summary with two additional elements. The first is a short section showing how parents can help, which often includes the gathering of resources, visits to the library with their child, helping with school trips, or assisting with specific activities. The second is a short statement saying that parents seeking further information on any aspect of the curriculum can request this from the school office. In this way the school is combining its curriculum information with specific requests for help from parents and is helpfully letting parents know that more information is required if it is needed. Parents are being made to feel involved, informed and welcome.

Many secondary schools could do well to follow this example. Because children are older and generally travel greater distances to their secondary school doesn't mean that their parents are less interested in what they do. Most parents would, sadly, be surprised if their sons and daughters brought home from secondary school a termly sheet outlining the main course objectives for each subject to be studied that term together with a request to support their son or daughter in their learning through providing specific resources or opportunities. However, as we move into an era of home–school agreements and contracts, smart schools will examine ways of making those contracts meaningful. These might mean spending a little time each term sharing information in return for an expectation of continued support from informed parents.

Obtaining comparative statistics

Thankfully, we are moving away from a situation where children had to protected so much from the concept of failure that their parents were denied any meaningful information on their child's or their child's school's performance when compared with the performance of other children or other schools. There were, justifiably, problems with labelling children as the worst in the class. Equally so, children labelled as the best in the class often had unreasonable pressure put on them to maintain that status. However, did it really make a difference?

Many successful adults, including ones heading the fastest-growing companies of the late 1990s admit to being 'a bit of a dunce' at school. Whatever the rights or wrongs of this situation, it could have not been any worse than the levelling of children under the guise of equality and almost complete denial of differences in ability that featured in many schools in the late sixties and seventies. This period was truly a 'dark age' in education, when the talents and capabilities of many children in many schools must have been wasted, either because they were not recognized or because they were not encouraged to improve.

Schools have made enormous progress in developing and sharing information in a useful and sensitive way, but there is still a long way to go. Technology helps, and will continue to help in the future. It is possible for a parent to be given a graphical indication of how well their child is attaining against the class average, against the LEA 'like school', or against the national average. Combine this with the previous year's similar information and the parent is being given a truly accurate and informative indication of a child's progress and attainment. Combined with good reporting and efficient communication systems, the information should enable parents to feel happy that the school is doing all it can to keep them informed of their child's progress and behaviour.

Handling complaints

All schools are going to receive complaints at some time or other. There are many ways of handling complaints.

One way is to ignore them, usually done by paying lip service to the nature of the complaint and instituting a half-hearted and inconclusive review of the 'circumstances'. Often the concerned parent gets fed up with asking for an outcome and loses interest. The problem with this approach is that it encourages parents to develop the complaint and second-guess the school's response in discussions with other parents at the school gates before and after school. Problems not resolved at an early stage can become much larger issues than they really are and can end up involving far more people than they need to.

A much more constructive approach is to hear the complaint as soon as possible, although absolutely immediate responses are rarely wise. It is better to commit to writing within a couple of days, and then saying how the school will resolve the problem. A well-crafted letter will outline the parent's concerns and state how the school intends to investigate those concerns. It will also give a response date. Of course, complaints regarding the school's handling of the health or safety of an individual child may have to be dealt with more quickly – often immediately in these cases – but a letter acknowledging the complaint and saying how it has been dealt with will still be gratefully received.

Some schools appear afraid of admitting that they have made a mistake. Trying to get an apology from a school over an incident when the school has clearly acted wrongly can sometimes be like trying to get blood out of a stone. This type of approach presents headteachers as arrogant and uncaring and does very little to engender a sense of trust and confidence in the school's ability to care properly for its children. If a hi-fi system breaks down, one well-known retailer usually apologizes straight away, sorts out the problem, and doesn't question the handling of the equipment by the customer unless it has quite clearly been mistreated. A school would do well to follow that example. Like service users or customers everywhere, parents appreciate courtesy, responsiveness, fairness and a high quality of information. The schools that are providing this are truly serving their communities well and setting examples to their pupils in the way they treat others.

Summary

Building a better school requires that a solid and constructive partnership is built with parents. Home–school contracts and homework agreements will mean nothing if there is not a serious intent behind these agreements to make them work. Making them work, as with any agreement, requires mechanisms that allow concerns to be shared early and effectively. It requires trust, and trust in a relationship usually comes with openness. Openness in education is about sharing information and being accessible.

Parents have a right to a range of basic information regarding their children's education. Table 6.4 summarizes that information and shows where and how schools could make it available.

CHAPTER SUMMARY

The information that governors, managers, teachers and parents each need to know to satisfy themselves that a school is succeeding has been identified in this chapter. Furthermore, the importance and relevance of such information as performance data and how it can be presented to, and considered by, each group to satisfy their needs is described.

The main points arising are as follows:

- ❑ For governors, matters relating to personnel, training, resources and accommodation are highlighted. General pupil information, school development, policy planning, and school successes and initiatives are also included.
- ❑ The headteacher needs to know how well teachers are teaching, whether their planning is adequate and relevant, whether teachers' assessment records are up-to-date and useful, what the training needs of staff are, how well resources such as the library or information resources facility are used, how effective are the arrangements for pupils' pastoral care and social development, and whether the arrangements for behaviour management are working.
- ❑ The teachers need to know how effective their teaching is, how relevant their planning is, what their pupils have learnt, what exactly they *need* to learn, and whether the pupils are making progress.
- ❑ Parents need to know whether their children are making progress and whether that progress is appropriate for their capabilities. They need to know whether teachers have concerns over their children's performance or behaviour. They need to know, in outline, what teachers are going to teach each term, and they need comparative performance information in a format they can understand so that they can see how well their children and the school are performing against others.

Table 6.4 *A summary of what parents should know and be told*

WHAT PARENTS NEED TO KNOW	HOW THEY CAN BE INFORMED
Parents need to know that their child is making progress and that progress is appropriate for the child's capabilities	Regular reports containing information about progress rather than simply attainment. Meetings to discuss progress with parents on demand as well as scheduled throughout the school year
Parents need to know whether teachers have concerns over their child's performance or behaviour	Effective systems within the school that bring poor performance or behaviour to the attention of appropriate managers quickly. Effective systems for contacting parents and arranging an early meeting at which the problem can be discussed and strategies addressing it can be identified and agreed
Parents need to know, in outline, what teachers are going to teach each term	Useful information that can help parents become more involved in supporting their child's learning either at home or through helping in school
Parents need comparative performance information in a format that they can understand so that they can see how well their children and the school are performing against others	Charts showing the child's performance against the class average, the school average, the national average, and against the average attained by children in 'like schools'. Charts showing the child's performance since the previous reporting period
Parents need to know how they can contact the school, who they should contact and when	Complaints procedures that show that the school values parental contact and makes it easy for parents to express their concerns. Strategies for handling parental complaints efficiently and effectively and for keeping parents informed of progress in any investigation. A preparedness to say 'sorry' if the school has acted unreasonably

7

WHAT TO DO WHEN THINGS GO WRONG

Schools are complex enough places to manage on a day-by-day basis when everything appears to be running smoothly. However, even in the best-run schools things will not always go to plan. The difference is that the best-run schools are able to spot the potential problems early and do something about them before they get really bad. It is unfortunate, as a corollary, that the schools already constrained by weaker management will probably be hit harder than others when problems arise.

In this chapter, consideration is given to when and how to act if difficulties with teachers, managers or governors are identified.

THE TEACHER IN DIFFICULTY

Everybody experiences moments in their working lives when things do not go as well as they should have done. Sometimes this can be despite good advanced planning and preparation that, on another day, would have ensured success in the task being addressed. Reasons for failure can be many, and can often be unconnected with the competencies of the individual or with the work or task being addressed.

A teacher's lot . . .

A career in education means ever-increasing demands for those who work in this field, whether they be teachers, school managers, governors (who, although technically not employed in the education sector, undertake what is undoubtedly becoming highly skilled voluntary work), or even education officers and inspectors. Teachers are particularly hard-hit as they are in the front line of educational reform. They above all others know the true impact of phrases such as 'from policy to practice', particularly when changes in policy and innovative new policies have been coming thick and fast for the last 10 years or more and show no sign of abating.

When the National Curriculum was introduced, teachers could be forgiven for thinking that the whole purpose of such an initiative was to give consistency and stability in the classroom for years to come. When Local Management of Schools was introduced, headteachers could have been forgiven for thinking that greater freedom meant less interference from outside bodies such as their local education authorities. Well, the National Curriculum has been revised, slimmed down and is shortly to be reviewed again. In addition, we have the Literacy Hour and the Numeracy Hour.

Each one of these changes has meant significantly more work for teachers as they rewrite their schemes of work, adjust their lesson planning and assessment practices, and become familiar with new professional vocabularies and ways of working. Those who believe that the outcome of the target-setting process in 2002, when schools and LEAs achieve (or fail to achieve) their first long-term literacy and numeracy targets, will not lead to further change are not living in the real world! There will inevitably be another round of changes to the English and mathematics curriculum. In addition, try as it may, the education establishment has still not got the curricula for design technology or information and communication technology right. And what happened to science by the way? Can a subject as important as science to the future of our society really be left on the back burner for so long? Expect more changes.

As far as school managers are concerned, have local education authorities really reduced the amount of paperwork they send out? There is some justification for answering this question in the positive, but unfortunately any impression of actual reduction has been totally distorted by the amount of paperwork received by schools from the local education authority as an unavoidable consequence of national initiatives outside their control or from the DfEE.

Some stress is good for you?

Against the background of change just described, teachers still teach and headteachers still manage. Is it surprising, though, that under such pressure they find that things go wrong from time to time? Pressure is an almost unavoidable circumstance in today's working environment. No matter what profession an individual is in, there is almost always pressure. Stress management is a thriving industry, and good employers are aware of the damage that a seriously stressed workforce can inflict on their business.

There is, of course, a difference between pressure and stress. The problem is that most individuals only find this out when they have experienced both. Pressure can be fun; there is often an adrenaline rush and a feeling of excitement and high energy. Knowing how to recognize this and control it is essential if the next stage of fear and exhaustion is to be kept at bay. It is only usually in retrospect that people realize that healthy pressure has turned into stress, and often the damage is then done to their working lives, their personal lives and their employer's business.

> One of the starkest manifestations of this tension between pressure and stress was observed by an inspector in a class at a college in the north of England where adults were being taught an ancient Eastern technique for relaxation. The teacher arrived late, having not been told that the inspector was coming to observe the class and the class having been moved to another room without her knowledge. These events caused the teacher to demonstrate a most expressive, and graphically Western, mixture of frustration, annoyance and hostility towards the inspector, and all in front of the students! The calm and peaceful atmosphere that had dominated in the minutes before the tutor's arrival, as students prepared themselves by undertaking meditative and mind-focusing exercises, was completed shattered by the inappropriate response to an unforeseen situation by the very individual in the room who should have been able to handle it.

Stress strikes in the most unusual circumstances and in this instance the poor tutor had not recognized the symptoms early enough. The tutor had all the techniques at her disposal to counter these symptoms and to deal with the situation rationally. Sometimes stress can be bigger than we realize!

Teachers who have a bad day despite the good planning and preparation that has served them well on previous days are not failing teachers. They should not think that they are, and poor performance on a one-off basis should not be seriously challenged. As long as the teacher is aware that something has gone wrong and is able to determine why, then in these instances there really is no problem. Inspectors are often asked by teachers if they would accept that the lesson just observed was not up to their usual high standard. An inspector can usually tell why a lesson is not going particularly well, and it is normally easy to spot the signs of a good teacher who is having a bad day. The clues, as we have already seen, are in the pupils' work, the teacher's planning, the atmosphere and relationships in the classroom, the quality of the environment, and the underlying features in the quality of teaching.

Teachers who are feeling stressed often feel that they are doing a bad job but, although stress can affect performance in the classroom, the stress itself is the issue that needs to be resolved.

Interpreting the symptoms

One of the essential skills of a good manager is recognizing when declining or poor performance is due to a skills deficit in new circumstances, personal or unforeseen circumstances that affect a teacher's performance, or an attitude problem to new or changing circumstances.

The skills deficit is often the easiest to identify from a manager's point of view, although sometimes external help is required to spot it. New circumstances or innovations often require new techniques and a good self-review programme within the school, supported by a comprehensive and adequately resourced professional development programme, should help most managers identify any deficiencies in the skills of their staff and take steps to address them.

Changes in personal circumstances can sometimes affect an individual's performance. We are not talking about the one-off situation where a teacher may feel unwell on a particular day but the major crises in somebody's personal life that eclipse work priorities and have a sustained knock-on effect on the quality of their work. Many managers will deal sympathetically with situations such as these and ensure that adequate support is offered. This should allow the employer to overcome the short-term acute difficulties and find a practical perspective in the relationship between their changed circumstances and their working commitments.

Sometimes, though, jobs change beyond the capabilities of the individuals to continue doing them, and sometimes there are people who cannot reconcile their changed personal circumstances with their professional commitments. In these cases, help has to be given to the individual in moving them to a new position or situation or even towards retirement. Where the individual accepts this and the employer has been fair and has given appropriate support, there need be no difficulty in working out an acceptable solution. This solution may include time and support to allow a change of job or career or a negotiated settlement package.

Considering formal proceedings

Some difficulties are caused by a teacher's unwillingness to accept that there is a problem that needs addressing. In much the same way that a weak manager must take ownership and responsibility for his or her own shortcomings before these can be addressed, the same applies to a weak teacher. The most difficult teacher to help is the teacher who ignores advice, who constantly challenges advice, who is always defensive, and who has a reason or response for every criticism or comment that may be made about his or her work and who will not accept that there are legitimate concerns about the professional performance being demonstrated. This is often the teacher who will blame management for the circumstances that cause the weak performance; or it may be that the teacher will blame pupils for not wanting to learn or for being disruptive. Under these circumstances a very specific set of rules ought to be employed by management and a very specific path followed, on the understanding that either performance will improve or the teacher will face penalties – even job loss.

Schools nowadays are required to have, by law, grievance and disciplinary procedures. Staff in many schools are wholly unfamiliar with them, as indeed are many managers and governors! For much of the time, most schools have more pressing things to prioritize onto their staff meeting agendas than examination of and discussion of the school's grievance and disciplinary procedure. In small schools particularly, headteachers don't really believe that they will ever need to use them and teachers can never see themselves in the position where disciplinary procedures would be being used against them. Whilst the governing body may have approved or adopted such procedures and the personnel group may have been through them in detail, a reference to them at a staff meeting and a quick read of the main headings will be as much as the general staff will have probably seen or wanted to see.

It is, though, often the case that when a potentially serious problem arises, such disciplinary procedures are referred to far too late. Often a period of months, or sometimes even years, when management has been aware of a problem, will have been allowed to elapse before disciplinary action is taken. This makes the use of disciplinary procedures all the more difficult. Many of the steps outlined in the early informal stages of these procedures, were they to have been implemented at an earlier date, could have helped resolve the difficulties much earlier and with a greater chance of improving the employee's performance rather than moving towards dismissal. Of course, it is easy with hindsight for a governing body, a headteacher or a local authority inspector to say that formal procedures should have been used earlier. Usually the best and most sensitive action at the time was taken by those responsible in the belief that it had the best chance of success whist being fair to all parties concerned. When this action doesn't work, it really isn't much help to wish that tougher or more formal action had been followed. But there are occasions when managers know of poor performance and prefer to play a waiting game, hoping that the employee will leave before they have to be formally dealt with.

Keeping accurate and up-to-date records

In every case there is a point – and that point is as soon as a problem is identified – where written records need to be kept and where concerns need to be shared with the teacher involved. Good self-review procedures will mean that managers have written records of all lesson observations, good and bad, and associated feedback notes on file, together with notes of professional discussions. Good practice will also have included targets being agreed, set and recorded after each observation and each interview for every teacher in the school. All this information will be helpful in determining the depth of any problem and will equally be fair to teachers through recognizing that there may be strengths as well as weaknesses in their performance.

When a concern arises, the process for addressing that concern will initially be no different from any other teacher's lesson observation as described above. Targets will be agreed and followed up. The difference in this case, however, will be that progress against the targets will be followed up sooner by a series of observations and feedback sessions. This is a very similar practice to what exists in the early, informal stages of most schools' grievance and disciplinary procedures. It is an easy operation to move from good self-review practice into informal stages of disciplinary procedures if it is necessary to do so.

If that happens, the teacher must know exactly what is happening and the disciplinary procedures must be followed to the letter. The reasons for taking this action must be clear to all. Simply put, that normal levels of support and advice have failed to help the teacher resolve difficulties; these difficulties are resulting in incompetent performance in the classroom; and the poor levels of performance are seriously damaging pupils' quality of education, the progress that those pupils are making and the standards they are achieving. It must be made clear to the teacher concerned that if the stated remedial measures are not successful in raising the teacher's performance to an acceptable level, then more serious action will follow.

The informal stages of a disciplinary procedure where there are concerns about a teacher's performance must give the teacher a clear indication of the concerns and a very specific expectation of what should be put right in order for the concerns to be alleviated. For example, a teacher may be having difficulty controlling his or her class and managers may also have noted that standards in that class are lower than elsewhere or that an examination of pupils' previous attainment shows that pupils are underperforming. Initially, concerns may have arisen because of comments or complaints from parents. In schools where there is a good system for self-review, managers will have picked up the class management difficulties. Where the self-review systems involve regular sampling of pupils' work, managers will also have picked up a decline in progress, in turn affecting attainment.

In many instances where advisers have assisted schools in addressing such issues, poor planning has been a major contributory factor. It is not the amount of planning that is at issue but the understanding of the learning process as far as it pertains to the children in the teacher's class at that particular time. In Chapter 3 we examined thoroughly the need for accurate and meaningful Learning Outcomes. If the teacher knows what it is that the pupils need to learn, then planning for progress and high attainment is much easier.

Sometimes, class control issues arise because a teacher is inconsistent in dealing with different children in similar situations or does not set clear expectations for behaviour that are understood by all pupils. Often, ignorance by the teacher of the school's practices over rewards and sanctions are contributory factors in these situations because this can make pupils feel less secure as they are treated differently in different classrooms.

Target setting and review

After establishment of the issues that need resolving – for instance, in the above example it is accurate planning and behaviour management – then targets can be set to help the teacher resolve these problems. Support is essential. The setting of targets on their own without support is unhelpful and unfair, for it leaves the teacher feeling accused and isolated. If the problem is shared, the solution should be shared too. Management must show a commitment to helping the teacher in difficulty by providing support to help that teacher overcome his difficulties.

In these instances, the support for assisting in planning may include the following:

- ❑ examination of exemplar planning in the school;
- ❑ assistance in using assessment to inform planning;
- ❑ assistance in writing meaningful and accurate learning objectives;
- ❑ discussion of teaching strategies that could be used to support the learning objectives;

- ❑ review of lessons taught with monitors' comments, where available, and discussions changes to make for future lessons.

Further, it may well be the case that planning is weak because the teacher does not fully understand the subject matter, in which case the following may also be required and should be provided:

- ❑ professional development to update the teacher's subject knowledge;
- ❑ professional development to update the teacher's knowledge of assessment in the subject.

The support for assisting in behaviour management may include:

- ❑ discussion with the teacher of the school's reward and sanction strategies;
- ❑ discussion with the teacher of the points in a particular lesson where problems may have occurred;
- ❑ identification of different strategies:
  - ❑ lowering the teacher's voice instead of raising it when a class begins to get noisy;
  - ❑ using silence and a good 'look' to attract attention and restore order rather than shouting over any noise that the class may be making;
  - ❑ never ignoring groups of children who may be talking in the hope that if you carry on teaching they will eventually shut up – they never do;
  - ❑ directing questions, as a matter of course in the lesson, to those pupils not paying attention and helping them get 'back on board';
  - ❑ using humour (but not sarcasm) to resolve a potentially confrontational situation;
- ❑ observing a more experienced teacher handling the same or a similar class and discussing everything about that lesson, including the planning and behaviour management strategies;
- ❑ attendance at carefully chosen courses if this is appropriate.

Improvement in performance should not be expected overnight, but it is reasonable to expect that if a teacher can begin to acquire new skills, for example in planning and behaviour management, then the application of these skills will begin to bear fruit in the classroom. For this reason, when targets are set it is vital that success criteria, to be achieved within a given time frame, are agreed at the same time. It would be unreasonable to expect a teacher's performance in the classroom to have all traces of weakness removed in a few weeks, but it would be quite realistic to expect that, in the space of half a term for example, the following outcomes could have been achieved:

- ❑ Planning has improved to the point where the teacher's understanding of the subject is clearly demonstrated.
- ❑ Learning objectives are being clearly identified for pupils and are linked to previous assessment outcomes.
- ❑ A range of different teaching strategies is being used to support different types of activity and different types of learning.
- ❑ Instances of failed lessons are decreasing.
- ❑ Pupils' work is showing more consistent improvement.
- ❑ Lessons are clearly structured and flow better.
- ❑ Lessons finish properly and on time.
- ❑ There are less interruptions to pupils' learning or to the teacher's teaching.
- ❑ There are fewer crises in the classroom and less confrontation.

Regular observations should take place, perhaps fortnightly, during this period and help should be given to the teacher throughout on planning and classroom management. At the end of the period, it may be clear that performance is improving and that the support can be continued but gradually withdrawn once management is happy that the issues have been resolved. If the issues remain unresolved, however, then targets may need to be reviewed, re-set and more rigorously followed up within the next stage in the school's disciplinary procedures. What is unsatisfactory is for the problem to be left unresolved and for the support to fade away before the next stage is invoked.

Figure 7.1 shows the target setting and review process for the above example, as applied to a fictitious teacher named 'Mark Wright'. Consideration of progress, including discussion with the teacher of evidence from observations and other sources against each of the success criteria, will help shape the targets for the next stage of the process.

Summary

In conclusion, it is important that weaknesses are identified as soon as they begin to emerge. For experienced and long-serving staff, this may be after the introduction of a new initiative or major change in the school. For staff who are new to the school – experienced or otherwise – this may be shortly after appointment. It is important for management to be able to see the difference between a temporary and explainable drop in a teacher's performance and an unacceptable decline in that performance due to a skills deficiency or poor attitude.

A management response must include identification of the problems with the teacher concerned and agreed targets, with sufficient support

Agreed performance targets for: Mark Wright			
Monitored by headteacher/LEA Inspector **Target date –** (6 weeks)			
Target	**Action**	**Support**	**Success criteria**
To improve lesson planning and content to ensure that the needs and interests of children are accurately met	Ensure that planning includes: appropriate lesson content for all ability groups; appropriate learning objectives; a range of appropriate teaching strategies	Examination of exemplar planning in the school Assistance in using assessment to inform planning Assistance in writing meaningful and accurate Learning Objectives Discussion of teaching strategies that could be used to support the Learning Objectives review of lessons taught and changes made for the next time Professional development as necessary	Planning that meets the school's expected standards Clear learning objectives Appropriate teaching strategies identified Pupils engaged during lesson Evidence of attainment through pupils' work
To improve class control so that learning can take place uninterrupted by poor behaviour	Ensure full awareness of school rewards and sanctions strategy Develop a repertoire of behaviour-management strategies	Discussion of the school's rewards and sanctions policy Discussion of observed lessons Identification of different control strategies Observation of other teachers Professional development, if appropriate	Behaviour management is consistent and in line with the school's policies A range of behaviour-management strategies is used Interruptions to lessons caused by poor behaviour or behaviour-management are decreasing
Summary of notes from observation noting performance against success criteria			
Week 2			
Week 4			
Week 6			
Recommendation			

Figure 7.1 *Target-setting and review process for a fictitious failing teacher*

and a realistic enough timescale to make the targets achievable. Management must be prepared to recognize progress and to continue support if it is reasonable to do so. Management must also be prepared to take more formal and tougher action without hesitation if the initial programme of support and target setting does not help the teacher meet the success criteria.

THE HEADTEACHER IN DIFFICULTY

Probably the most difficult task that a governing body ever has to undertake is that of dealing with a headteacher who is clearly failing in the post. We have explored elsewhere the relationship between the headteacher and the governing body of a school and noted how a successful relationship is reliant on trust, a clear understanding of the roles and responsibilities of each party, and a free flow of information between the two parties. One of the most difficult aspects of a situation where a headteacher's performance is below standard is that the headteacher will usually do everything possible to hide the difficulties from the governing body or play down the scope of the problem.

Change and the established headteacher

Schools can run into difficulties with their headteachers for a number of reasons. Sometimes a headteacher has been in position a long time. Circumstances may change and the headteacher may not be able to handle the effects of these changes. For example, the school may change through an increase in size or because it begins to draw pupils from a different catchment area. The socio-economic factors within a catchment area may change over time as economic fortunes within a local community change. Areas can decline rapidly because of the permanent closure of local industry; conversely, areas can improve because of the 'gentrification' of old Victorian properties. Both sets of circumstances have had implications for the schools within such communities.

In the last 10 years, headteachers have been expected to become managers *in* education rather than managers *of* education. They have become ever more isolated from the classroom and, with each successive national initiative or change, they are becoming further removed from the realities of day-to-day teaching. Very few headteachers in primary schools will have taught the national Literacy Hour or Numeracy Hour, but they will all be expected to manage it. And it should be remembered that schools inspectors and education officers, together with teacher trainers, have long been accused of being remote and removed in this way.

Nevertheless, there are those who have a clear understanding of the education process, and who have an up-to-date knowledge of education initiatives and techniques. They are prepared to consider their time with teachers in the classroom as much as a learning experience for themselves as a learning experience for the teacher they are observing. Once appointed to the position of headteacher, these people have a responsibility not to impose out-of-date ideologies on those remaining in the classroom. Accepting that good practice is always evolving as society, the curriculum and the expectations of children change, and accepting that what works for one may not work for another, these individuals are in a much stronger position to manage education and share good practice than those who, once they have achieved the status of headteacher, lock themselves away in their offices, and espouse only their own 'proven' techniques.

The weak appointment

Often a failing headteacher is a failing headteacher from the day of appointment. Despite interviews that may last a day or more and that may include in-tray exercises, simulations and presentations, the slick interviewee with a good eye for presentation may lack that essential touch of being able to relate to colleagues, children or parents, or of being able to make a difficult decision decisively, and this can often be hidden in traditional headteacher interviewing techniques.

Within months of an appointment, governors may begin to realize that their expectations are not being met as parents complain of inaccessibility, or long-serving and highly respected class teachers begin to leave for no apparent reason. There is a real problem for governors here. Initially, no one is quite sure whether the parental complaints are those of 'troublemakers' who would have complained anyway, and the experienced teachers who leave are usually far too polite to give their real reasons for departure. Some good teachers have been known to leave a declining school rather than 'blow the whistle' because the fear of not being believed and the fear of jeopardizing their chance of a decent reference in the future have together been a major consideration for them. In other instances teachers close themselves off from the rest of a failing school because they do not want the stress, upset and discomfort of causing difficulty for another person; they do not wish to move, or for family reasons cannot move, and simply want to teach in a professional and supportive environment. Furthermore, it is the governors, of course, who made the appointment in the first place, and it is not easy to admit that you have got something so seriously wrong as the appointment of your headteacher – especially when any evidence is so hard to corroborate.

It will nonetheless dawn on such a governing body sooner or later that there is a problem and that something must be done to resolve it. There is now a requirement for local education authorities to inform a governing body if they have concerns about a school or its management. This automatically triggers the creation of an action plan that the governing body must impose in order to rectify the problem. In the past, local education authorities have sometimes been reluctant to share too much with governing bodies because if the governing body rejected the concerns wrongly, for whatever reason – and some have done exactly this – it has become even more difficult to sort the problem out with the errant headteacher. Situations like this have been known to drag on for years, potentially damaging the education and future of hundreds of innocent children passing through the school.

Early identification

So what is a governing body to do if they have concerns about their headteacher? How are they supposed to know whether or not they should be concerned? Unlike the relationship between headteacher and class teacher, there is no system obviously parallel to that of regular lesson observations that may alert them to shortcomings in the way that a headteacher could be alerted through seeing weak teaching. The important consideration for a governing body to have in mind is that they must avoid being taken by surprise in the event of anything going wrong. Far better that they discover for themselves that there is a problem at an early stage than be told by a local education authority, or other authority, that there is a problem. If the latter happens, then in the eyes of the authority the governing body is usually seen as part of the problem because it did not have the systems to pick the concerns up itself – or if it did, it has not acted on those concerns.

If a governing body knows its headteacher well and there are systems for discussing any issues that are facing the head or the school each week, then both governors and headteacher have a greatly increased chance of preventing any minor difficulties becoming major obstacles. In a culture where a headteacher is comfortable in sharing issues with key governors and where these governors encourage this as a sign of strength and not an acknowledgement of weakness on the part of the headteacher, longer-term problems are less likely to arise.

Take, for example, the following account of parents who wish to discuss a concern with their child's headteacher.

> The parent struggles to get an appointment with the headteacher, who is nervous about 'confronting' this particular person. Even before a time for the meeting is eventually arranged, word has got round the parent community that the headteacher is evasive and not willing to listen to parents. Once in the meeting, the headteacher refuses to accept any criticism of the school, despite compelling evidence to justify the complaint, and even suggests that if the parent is not happy then his children should be transferred elsewhere.
>
> Other parents hear about this treatment and begin to express open concern, which reaches the ears of prospective parents, who begin to look at schools elsewhere. Governors hear about the parents' concerns and eventually, as more instances of this nature occur, some governors go to see the headteacher. However, they do not go with a solution to the headteacher's problems but with a complaint of their own about how parents are losing confidence in the school. Because of the headteacher's evasiveness with the parents and denial of the concerns raised by parents, the headteacher has already lost the support of the very people who could have been instrumental in helping resolve the problems.

Now compare that situation with this one that was found in a different school.

> A headteacher receives a parental complaint. He invites the parent in to school immediately and hears the parent's concerns in detail. He makes no comment and offers no defence. Having clarified with the parent the full nature and circumstances of the complaint, he promises to have the whole issue investigated and to speak to the parent again within a specified time.
>
> The headteacher may or may not have known that the parent's complaints were groundless. However, in the parents eyes there were grounds for a complaint and the headteacher is wise to recognize this and have the issue investigated rather than dismiss

the parent outright. This headteacher is already streets ahead of his unfortunate colleague in terms of public credibility, but the best is yet to come. When the parent rejects the headteacher's initial findings, as parents with problems can often do, and wishes to press the complaint further, the headteacher arranges for the parent to see governors who, as outlined in the school's published complaints procedures, are the next point of contact for dissatisfied parents.

Other parents are impressed at the even-handedness of this approach and the headteacher is confident enough about his interpretation of events to trust his governors to handle the complaint fairly. Even if they find contrary to his own conclusions, his relationship with his governors is strong enough to withstand this. What matters is that the parent has been treated fairly and the headteacher has played his part in ensuring this through working with the governors.

The second headteacher could only do this because of two considerations. The first was that the school had a publicized and consistent response to all parental complaints. The second was that the headteacher had a strong relationship with his governors based on trust and professional respect. The same headteacher had been known to discuss difficult issues with his chair of governors before coming to a decision or conclusion; and he had been open to advice and sought it when he felt this necessary. The first headteacher never had any difficult issues to discuss with anyone and certainly did not think that he would need anyone else's advice.

Making sure that the chair of governors, or the vice-chair, has regular contact with the headteacher to discuss issues and to develop mutual respect and understanding is the first action that a governing body can take to support its headteacher. If a headteacher rejects this kind of relationship, then that in itself may be a reason for concern: headteachers who have had to leave their positions because of a dispute have often not had this kind of positive relationship with any of their governors.

The occurrence of regular meetings between governors and a headteacher makes it easier for governors to raise issues informally in the other direction. If concerns from parents are reaching the ears of governors direct, then they can be brought up and discussed in an unthreatening manner in such a meeting. If no regular forum exists, it will mean calling a special meeting to discuss a parental complaint. This will immediately make the headteacher feel threatened and defensive and will reduce the chances of a successful outcome.

Dealing with ongoing difficulties

The real difficulty comes in these instances where the headteacher fails to convince the governors that sufficient action has been taken and the issue in question continues to deteriorate. If informal discussion and support (perhaps through governor or outside specialist contribution) has failed to resolve the problem, then more formal solutions will have to be initiated.

Once again we turn to the school's grievance and disciplinary procedures. Often these are not phrased in such a way as to be easily interpreted for use with a headteacher and the advice of an education personnel specialist will be needed to ensure fairness and procedural correctness. It is wise at this stage, in any case, for governors to take advice from an appropriate authority in implementing procedures of this kind.

Just as with the procedures for teachers, the early and relatively informal stages can involve clear identification of the problem with the headteacher, the agreement of targets, and the provision of support to help meet these targets. Professional assistance may well be useful at this stage and throughout the process. Moving from informal stages to formal stages should never be taken lightly, whether for teachers or headteachers, and clear and measurable criteria need to be agreed so that all parties can see whether the rationale for such a move is necessary or not.

Examples of measurable success criteria pertaining to the previously mentioned examples are as follows:

- ❑ A policy agreed with the governors for meeting parents and handling complaints has been fully implemented.
- ❑ Procedures for keeping parents informed of progress of any investigation have been agreed and all meetings and action noted.
- ❑ Governors are satisfied that set procedures are being followed.
- ❑ Key governor(s) are being informed at an early stage of emergent difficulties likely to require their attention.
- ❑ Parental complaints about accessibility to the headteacher are decreasing.

It should be noted that it would probably be unreasonable, no matter how desirable, to have a reduction in parental complaints as a measurable success criteria. This is because, at this stage, the issue is how complaints are received and not whether or not they are justified.

Once formal stages are introduced, there is almost always an irrecoverable loss of confidence between headteacher and governors. This in itself can make future working relationships difficult to say the least, and can often result in an early departure by the headteacher or a long and acrimonious relationship that will eventually end in severance. However, a governing body that has been clear about the headteacher's strengths and weaknesses from the outset and that is operating from a position of knowledge, understanding and trust will handle such difficulties more effectively than a governing body to whom serious problems with their headteacher have come as a shock. Table 7.1 shows how movement through such a process can be managed.

Formal proceedings are always difficult to initiate, but if a school's regular monitoring and support practices are suitably structured then the progress through these to formal competence proceedings will have an easily understandable logic. The key word in the school's regular monitoring activities and in the informal proceedings described above is 'support': few people will object to a criticism if it has emanated from a recognized and fair procedure and if proper support is offered to help meet that criticism.

Summary

The principle to be followed in how a failing headteacher can be brought back on track is once again one of partnership. In this case it is the governing body helping the manager of the school to rectify a recognized problem by providing support. Phrased in these terms, there is nothing more to it than that. Ultimatums from previously uninvolved governors to resolve previously undisclosed issues will always provoke resentment and confrontation. Similarly the headteacher who will not share, or admit to the existence, of an issue is also following the path that could easily lead to confrontation.

Handled properly and at an early enough stage, the problem may not be a personal one with fault attributed to the headteacher. It may well be an institutional one because systems were not in place somewhere in the school to handle the problem that caused an original complaint. Handled improperly, these institutional shortcomings are forgotten and the big issue can easily become the headteacher, with serious consequences for the whole school as the difficulties are worked out.

THE GOVERNING BODY THAT CANNOT GOVERN

One of the strengths of educational reform in the last decade has been the shift of control and accountability from the centre to the school. This has meant increased influence and responsibilities for governors. The end of the Grant Maintained era will not affect this shift in the long term because

Table 7.1 *Review process for a failing headteacher*

	KEY INDIVIDUAL	ACTION	SUPPORT	OUTCOME
Initial identification	Chair/Vice-chair of Governors	Discuss problems. Establish facts	Other governor; LEA inspector; other authority officer	Problem agreed and solution identified and implemented
Problem persists/ deteriorates	Chair/Vice-chair LEA or other authority officer	Offer support. Agree targets for problem-resolution success. Agree success criteria	Professional support from LEA inspector/ officer or consultant	Success criteria met and problem resolved
Success criteria not met/problem persists	Chair/Vice-chair/LEA or other authority officer	Move to first stages of competency proceedings. Re-visit targets and success criteria and revise if necessary. Provide support	Personnel advice; professional support from LEA inspector/ officer or consultant	Targets and success criteria met
Targets not met	Disciplinary hearing	Follow formal stages of competency proceedings	Personnel advice; professional support from LEA inspector/ officer or consultant	Targets and success criteria met or formal proceedings continue

successive governments appear to be committed to maximum devolution to schools and minimum interference in schools by local education authorities.

However, one weakness of this shift of power and influence is that it assumes a consistency of performance from volunteer governors. This assumption is wholly unrealistic. Governors change – perhaps more frequently than the teaching force in their schools – and, as we have seen, new governors generally arrive untrained and unprepared for the role. There have been cases – and there are bound to be other cases in the future

– where a weak governing body becomes a liability to a school, particularly if the school has other weaknesses. This section examines what can be done to counter such a situation.

Independence, isolation or interference?

To some headteachers a quiet, uninvolved and totally trusting governing body of the 'one meeting a term and rubber-stamp the head's decisions' type is an absolute dream: power and control for the head without interference and accountability, or time to do the 'real' job without wasting any in endless meetings with 'people who need to have everything explained to them'. However, such governing bodies are becoming rarer and the headteachers who can successfully and properly manage a school without the support of a governing body are becoming very few and far between. Legislation through Education Acts and supporting Regulations (for example, the Education Act 1996) has made it quite clear that governors have specific legal responsibilities and accountabilities and there are some things that a headteacher cannot do unaided, such as setting the budget and appointing staff.

To other headteachers (the majority, fortunately) a quiet, uninvolved and totally trusting governing body is a liability. Total trust means that accountability is, in the eyes of the governors, delegated to the school management team. Yes, they may agree the budget having not understood it or contributed to its development. Yes, they may be happy that the headteacher appoints staff without governors present. However, should the school's budget position become unfavourable or should there be difficulties with a teacher's appointment, then the responsibility, as far they are concerned, rests with the headteacher rather than themselves. There are headteachers in this very difficult position who simply cannot get governors to understand their responsibilities or to undertake the tasks they are supposed to do. In small schools this puts unreasonable pressure on overworked headteachers to do the work of the governors as well. One other point needs to be made in these circumstances: with a weak governing body the danger is that individuals will become offended and leave the governing body; whereas this may give the opportunity to recruit more committed governors, there is the possibility that relationships will become strained with the remaining governors, making progress by the headteacher even harder.

There is another scenario. Some governing bodies are so immersed in their responsibilities to their own community and their care for the school that they fail to recognize when it is time to back off and let the headteacher get on with managing the school. The point when governing bodies begin to make, or interfere with, day-to-day decisions or practices means serious problems for the school. In these instances it can be very difficult for a headteacher to know how to handle the situation. These may well be the people, after all, who employed the headteacher in the first place. With an overactive governing body, confrontation may become a battle of wills and the headteacher, as the employee, is not going to win this battle alone.

Early recognition

Prevention is clearly the best course of action and at the outset the respective roles of the governing body and headteacher should have been clarified. However, new headteachers can find themselves appointed to schools with very weak governing bodies and may not have the experience to address and resolve their relationship during the 'honeymoon period'. The problem then deteriorates as the new headteacher begins to flounder, unsupported, under an unrealistic workload. Similarly, existing headteachers who may have had a little difficulty with some governors may find themselves facing an increasingly strong and interventionist governing body if the difficulties have not been properly resolved. Either way the headteacher needs help because, unless the problem can be sorted out at a very early stage, it may not be possible to sort it out without external help later on.

With weak governing bodies the problem is usually one of inexperience on the part of the governing body and unfamiliarity with their own responsibilities. If there is a willingness to become more involved and to take their responsibilities more seriously, the problem is relatively easy to resolve. If that willingness is not there, then there may be a need for the appropriate authority to offer training; and if that does not change the situation, then it may have to appoint new governors. The involvement of an external adviser at an early stage is essential if the problem is to be successfully resolved. The longer it is left, the longer it becomes a joint problem, whereby the governors can easily say, as a countercharge, that the headteacher does not keep them fully informed or allow them to appropriately exercise their responsibilities.

Seeking external help is not necessarily a sign of weakness. It is more a recognition that there is an extremely delicate situation that it would be appropriate to have the assistance of an impartial third party to resolve. An adviser is then the individual who brings the chair of governors and headteacher together to talk about their respective responsibilities and who offers or provides training for the governing body on recognizing its responsibilities, setting itself up to deal with them, and managing its workload. This 'learning together' approach, where headteacher and governors set up new systems as partners with the assistance and guidance of an external agent, is much more comfortable than trying to work things out unaided.

With overactive governing bodies, the problem is usually one of lack of confidence by the governing body in the abilities of the headteacher to perform to their standards. It may well be that there has never been a discussion at which expectations were agreed. What is likely is that the headteacher has never been given a proper chance to understand what, if any, concerns the governors have, nor therefore had the chance to rectify them along the lines discussed in the last section of this chapter. In such circumstances, governing bodies will be making decisions and taking actions that can potentially undermine the headteacher, making it even more difficult for the headteacher to operate successfully – for example, meeting staff outside the normal framework of committee meetings without the knowledge of the headteacher, or commissioning work to be done or resources to be ordered without discussing the matter with the headteacher. Other actions may include the imposition of new financial procedures or a new governing body structure without consultation with the headteacher. All these actions can and will appear threatening to a headteacher, particularly if there has already been some friction between the two parties. This can result in making any problems even more difficult to resolve.

Retrieving the situation

If the governors are not using clear and agreed procedures to get the performance they want from the headteacher and the headteacher feels that they are acting unreasonably, the only way to resolve this situation fairly is through the use of external advisers. It is important that both sides want a resolution to the difficulties. External advisers will be able to identify where the governors are acting illegally and where there are acting ill-advisedly; they will also be able to advise on how governors should deal with perceived shortcoming by the school's managers.

Pulling a situation back from the brink is a difficult and sensitive operation. Sometimes there cannot be a negotiated settlement between the parties concerned and someone has to go. There have been instances where the headteacher has left, others where governors have resigned, and a few where both the headteacher and several governors have left. While all this is going on, of course, the teachers are still trying to teach but the quality of education, and ultimately the progress and standards achieved by pupils, is almost bound to be affected.

A legacy of strength and success

Given that a built-in weakness of any governing body is the rate at which its membership changes, it is vitally important that systems and working practices such as those described in Chapter 5 are devised, documented and implemented. Such systems should be strong enough to be the constant factor that provides security to the variable nature of the governing-body membership. When a new chairman and several newly elected parent- and teacher-governors take over from a highly successful and established operation that has reached the end of its membership, well established systems and working practices will ensure that problems faced are minimized while new relationships are developed. Over time, of course, these new relationships will lead to changes in the systems and practices, but these will be evolutionary refinements rather than emergency alterations to inadequate systems.

Summary

It is essential that any difficulties with the governing body are resolved early. On appointment, new headteachers and governors should seek to establish clear working relationships and good lines of communication. If there are difficulties, then there should be procedures to deal with them, and if these procedures are not working or it is not possible to agree on a way forward, the headteacher should seek external advice from an appropriate authority.

CHAPTER SUMMARY

This chapter has aimed to give advice when certain difficult, but fundamental, relationships start to break down. The list below summarizes the main points:

- ❑ It is important to take immediate action when problems are identified.
- ❑ Schools that have procedures for supporting teachers in difficulty will have a greater chance of minimizing and resolving problems than schools that do not.
- ❑ Management needs to be confident in knowing when to move to formal disciplinary action.

- ❑ Governors need to know the techniques and procedures used to handle headteachers who are in difficulty. They need to know when to call in outside support and what kind of support is required. They also need to know how to handle a situation so as not unnecessarily to harm relationships between the headteacher and the governing body.
- ❑ Similarly, headteachers need to know how to identify and deal with problems caused by a failing governing body, including how to encourage a passive set of governors to take their legal responsibilities seriously and how to deflect a too-active set of governors from trying to manage the school on a day-to-day basis.

8

PLANNING FOR THE FUTURE: INSURANCE AGAINST FAILING

In this final chapter the essential foundations necessary for a successful school are considered. The appointment of the right staff – not just bodies in front of classes – is considered, as is the appointment of governors.

Schools should know their priorities and have a clear idea of how they are developing. This development should be supported through good financial planning and through judicious use of target setting for whole-school improvement. Knowing your own strengths and weaknesses and working on them together is the essential element in building a successful school.

APPOINTING THE RIGHT STAFF

In Chapter 2 we touched on the need for schools to have a staffing structure that is designed to meet the corporate aims of the institution. Behind such a structure should be a profile of skills, experience and personal attributes that are needed by staff in order to deliver its aims. The link between the required profile and the day-to-day work being carried out is meaningful job descriptions for each member of staff.

In this section we shall look in more detail at the required profile of skills and how this profile should be integrated with the development of an appropriate staffing structure, the formulation of job descriptions and, eventually, the appointment of staff.

Developing the staffing structure

Most organizations have a staffing structure, and schools are no different. At some point, even in the least well-organized of schools, there will have been some rationale in appointing a headteacher and deputy headteacher and in giving a number of teachers more money for carrying out a range of identified responsibilities. However, all organizations need to change and evolve to meet changing demands and circumstances.

Many organizations have developed, at some point or other, a 'shadow' staffing structure. This outlines the most desirable profile in terms of positions of responsibility and hierarchy that the organization needs in order to meet most effectively its current challenges. A second 'layer' of the shadow structure will indicate the skills, experience and personal attributes required to complete the profile. Schools may develop a 'shadow' structure as a result of new demands, a change in curriculum or a change in management style.

Few schools, however, have the strength, skills and commitment of a good management team to turn the shadow into substance without significant, and potentially damaging, upheaval. Instead, teachers remain in the positions they are in, or reluctantly accept 'new' positions that they are wholly unsuited to, and the full effect of the vision behind the creation of the shadow is put on hold until natural personnel changes allow the appropriate new skills to be put in place. The problem with this approach is that the pace of change in education has been so fast in recent years that new shadow structures have had to be developed before the old ones have been fully implemented. Schools in these circumstances indeed find themselves continually chasing shadows!

In a well-managed school, changes to the skills profile of the staff can be more easily and successfully made. Where staff are well motivated and management is seen as effective, decisive and reasonable, staff will respond well to training opportunities and the possibility of new positions of responsibility. Conversely, where staff are not well motivated and management is seen as weak or indecisive, staff will enter into new responsibilities with reluctance and without the commitment necessary to see the new job through successfully. These are the institutions that have least success in implementing change that leads to improvement.

Any significant change in the structure of a school that involves reorganizing responsibilities for the middle management is not going to work unless the senior management has credibility and is proven in managing its own responsibilities first. Before considering changes to the responsibilities of teachers, governors and senior managers must make sure that the role of the headteacher and deputy headteacher are clearly defined and understood.

So what makes a good staffing structure? Within a small primary school the management can afford to look at the profile of the whole school. Within a large comprehensive school, it may be possible to consider the profile of each department or faculty once the overall staffing structure for the whole school has been decided. In this case, the task is multidimensional. The management needs to decide on the range of skills, experience and attributes needed at senior-management level, at middle-management level (such as faculty head level) and then within each faculty. The model used can be the same at each level or, in the case of the small primary school, for the whole organization.

An integrated approach

The development of a good staffing structure through to filling posts with good-quality candidates is based on six integrated steps.

Step one: defining the important outcomes

The first step is to define the most important outcomes that are needed to make the school run successfully. This is another way of looking at the school's real priorities and should help to get into proportion a range of demands that are made on the 'management' profile of a school.

Take, for example, the organization of a performing-arts faculty in a large comprehensive school. The school first must decide the priority of the subject on the timetable and in the curriculum. The desired outcome may be that all pupils will experience dance, drama and music in Key Stage 3 and that all pupils will have the opportunity to study these subjects for GCSE and at post-16. The outcomes may be defined further; for example, apart from having the access to the courses in each key stage and at post-16 level as described above, the school may also decide that it wants to stage a major event in each art form every term and a major multidisciplinary event, such as a musical or opera, once each year. In addition, the school may want to raise its arts profile through an annual programme of external visits in each discipline and through a programme of visiting artists in residence. Table 8.1 analyses the desired outcomes in this example.

Step two: identifying the skills and experience profile required

A school with the above desired outcomes is clearly making a significant commitment to arts education. The size of the school will enable it to calculate how many teaching hours are needed in each subject. Assuming that this has been calculated to identify a need for up to 2.5 music teachers, 2.5 drama teachers and 1.5 dance teachers, the next step is to use this information to identify the skills and experience profile needed. Table 8.2 shows that analysis.

Step three: designing a staffing structure

The profile shown in Table 8.2 for our example could then be refined, taking into account management preferences for the subject skills of the co-ordinator. The resultant staffing structure could look like that shown in Figure 8.1.

Table 8.1 *Faculty of Performing Arts: desired outcomes*

KEY STAGE	ENTITLEMENT	ACCREDITATION	OPPORTUNITY	ENHANCEMENT
3	All pupils to have weekly timetabled lesson for one term each year in dance	None	In-house and occasional public dance productions	Opportunity to take part in visits to theatre. Work with visiting company/specialist
3	All pupils to have weekly timetabled lesson for all three years in music and drama	None	In-house and occasional public concerts and drama productions	Opportunity to visits to concerts and theatres. Work with visiting musicians and theatre companies
4	All pupils to have access to dance, music or drama options at end of Key Stage 3. Options to allow for two of the above to be studied simultaneously	GCSE in dance, music, drama or in combined arts. Non-vocational enrichment option	Annual performances in dance, music and drama	Termly visits to theatre or concert hall. Work with visiting dance specialists, theatre companies and musicians
Post-16	Appropriately qualified students to have access to next level of dance, music or drama education. Others to have access to lower levels of arts education	A level dance. BTec Perf.Arts. Non-vocational enrichment option	Termly performances open to public. Annual multi-disciplinary performance	Termly visits to theatre. Work with visiting dance specialists

Table 8.2 *Faculty of Performing Arts: skills and experience profile*

KEY STAGE	SUBJECT KNOWLEDGE	EXPERIENCE	ORGANIZATIONAL	MANAGEMENT
3	Specific. Competent in teaching National Curriculum music and in teaching dance and drama with regard to National Curriculum orders in PE and English. Knowledge of, competency in teaching second performing art desirable	**Music one** minimum three years **Music two** NQT considered **Dance** minimum three years **Drama one** minimum three years **Drama two** NQT considered **Music/drama three** NQT considered **0.5 drama/dance/ music** dependent on skills of co-ordinator	Music and drama. One teacher must be competent in staging concerts, organizing rehearsals, external visits and working with visiting artists The dance teacher should also have the above capabilities	One teacher should have the ability to oversee and co-ordinate the work of all three subjects throughout the school One teacher in each subject should have the ability to manage that subject within the whole performing arts structure
4	Specific. Good knowledge of GCSE requirements in specialist subject			
Post-16	Specific. Advanced knowledge of subject and of A level and BTec courses			

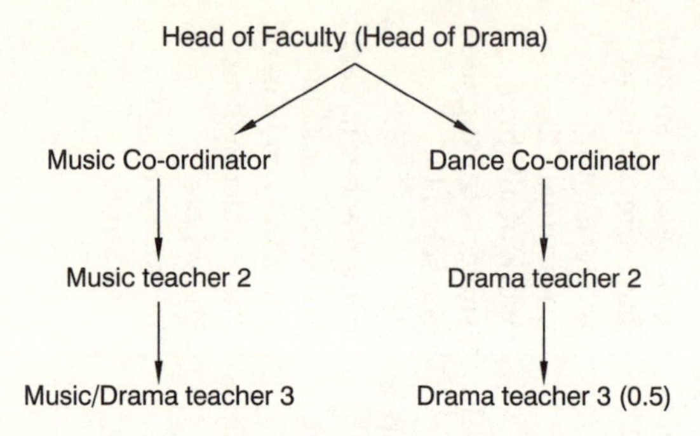

Figure 8.1 *Faculty of Performing Arts: possible staffing structure*

Step four: creating suitable job descriptions

The next stage, after decisions on the skills and experience profile and the staffing structure that the faculty needs, is to create suitable job descriptions for each individual post, ensuring that each of the identified skills and the desired experience profiles are represented appropriately within the job descriptions.

A job description for the post of Music Co-ordinator, for example, could take the form shown in Figure 8.2, drawing directly from the desired skills and experience profile already identified.

Step five: creating person specifications for the posts

If individuals are not in post and recruitment needs to take place, then a person specification also needs to be created. Again, drawing on the skills and experience profile already decided for the Music Co-ordinator position in our example, the person specification could take the form shown in Figure 8.3.

Step six: appointing the most suitable person

In appointing the person best fitted to the above specification, it is essential that the interview process is designed to test and, where possible, identify, the qualities inherent in the job description (see Figure 8.2) and set out explicitly in the person specification (see Figure 8.3). It is therefore worth examining each quality and deciding how best that quality can be tested or identified for each candidate for the post. This should give some indication as to the type of interview required to identify the most appropriate applicant.

Responsible to: Head of Faculty

Responsible for: Full and part-time music teachers

Grade: Two responsibility points

Key responsibilities:

- ❑ planning and maintaining the curriculum for music in Key Stage 3 in line with the requirements of the National Curriculum;
- ❑ deciding with colleagues and with management on appropriate courses in music for accreditation at GCSE and A level;
- ❑ deciding with colleagues the most appropriate combined GCSE arts course(s) and post-16 combined arts course;
- ❑ devising and contributing to the teaching of a non-vocational option at post-16;
- ❑ managing the peripatetic instrumental teaching timetables;
- ❑ monitoring, with the faculty head and senior management as required, the quality of teaching and learning in the department;
- ❑ ensuring that departmental and faculty assessment systems are maintained and monitored;
- ❑ ensuring that arrangements are made for weekly rehearsals of the school choirs and orchestra;
- ❑ co-ordinating the efforts of other non-music department staff in running music ensembles and contributing to the choir and orchestra rehearsals;
- ❑ ensuring the annual production of a music concert and the contribution of the music department to an annual multi-disciplinary arts event;
- ❑ monitoring the quality, condition and usage of resources and drawing up priorities for the replacement or repair of existing stock and the development of new resources;
- ❑ ensuring that a programme of termly visits to concerts is developed, suitable for each key stage and for post-16 students;
- ❑ ensuring that an annual project involving visiting musicians, supporting the aims of the curriculum and the department, is facilitated.

Figure 8.2 *Job description for post of Music Co-ordinator*

Such an analysis is given in Table 8.3 for our example.

The process of identifying required qualities for this particular position of Music Co-ordinator suggests that there are six separate elements (application form, references, audition, choir rehearsal, written exercise and interview) that need to be included in the selection process in order to assess the qualities of each candidate thoroughly.

Each of the six elements is described further below:

- ❑ *Application form*. This must ask the right questions. Consider carefully the information you need to know, with particular relevance to any gaps in career history, reasons for leaving previous jobs, and reasons for applying for this position. Employers may also want to ask for a letter, or paragraphs, seeking the candidates' views on specific issues relevant to the job being advertised.
- ❑ *References*. A reference request must be specific on subject-related questions in addition to issues such as health, punctuality, working with others, and other similar issues.
- ❑ *Audition*. An opportunity to hear the candidates perform (in this instance on the piano) is essential. The piece performed would be chosen by the candidate and would be of a length determined by the interview panel.

Experience:
- ❑ minimum of three years working in a successful music department;
- ❑ GCSE teaching experience required;
- ❑ A level teaching experience required;
- ❑ experience of performing arts courses desirable.

Qualifications:
- ❑ music graduate;
- ❑ qualified teacher status.

Skills:
- ❑ pianist with good accompanying skills;
- ❑ proven organizational skills;
- ❑ good presentational skills in staging concerts;
- ❑ ability to motivate other staff and pupils;
- ❑ ability to organize the work of visiting teachers and to liaise successfully with external agencies.

Knowledge:
- ❑ good subject knowledge at each key stage and post-16;
- ❑ knowledge of assessment-technique procedures at each key stage and post-16;
- ❑ good knowledge of concert repertoire for secondary pupils;
- ❑ knowledge of arts in education networks;
- ❑ awareness of how performing arts disciplines can work together successfully;
- ❑ knowledge of potential career paths for students and of opportunities for pupils to extend music experience out of school.

Attributes:
- ❑ ability to work with others as part of a team on curriculum issues and in the staging of arts events;
- ❑ ability to lead others, including those from other arts disciplines, on combined arts projects;
- ❑ confidence in performing and in conducting or directing the performance of others.

Figure 8.3 *Personal specification for post of Music Co-ordinator*

- ❑ *Choir rehearsal.* An opportunity to see each candidate rehearse a choir on a section of a piece selected by the panel is essential. The choir can be a combination of other candidates, staff and student/pupils. The rehearsal would be timed to a specific length – for example, 10 minutes.
- ❑ *Written exercise.* It may be useful in this case, but it is not essential. The candidates could be asked, as a pre-interview task, to describe the attributes of a well-organized music department working in the context of a performing-arts faculty.
- ❑ *Interview.* A formal interview is obviously necessary, but in this instance the presence of a music specialist is essential. It is easy for a musically trained teacher to appear a lot more competent than he or she actually is if the panel is ignorant of the levels of subject-specific skill, knowledge and competency required. If necessary, a school should bring a specialist adviser in for the interview, even if this costs something – it will be worth the money in the long run.

Each candidate's performance in all of the above elements needs to be assessed quickly by each interviewer. This is best done on a standard form (see Figure 8.4) that uses the headings and detail of the Person

Table 8.3 *How to identify personal qualities for a post: Music Co-ordinator*

QUALITY REQUIRED	IDENTIFICATION PROCESS
Experience:	
Minimum of three years working in a successful music department	Application form/references
GCSE teaching experience required	Application form/references
A level teaching experience required	Application form/references
Experience of performing arts courses desirable	Application form/interview
Qualifications:	
Music graduate	Application form
Qualified teacher status	Application form
Skills:	
Pianist with good accompanying skills	Audition/choir rehearsal
Proven organizational skills	Written exercise/interview
Good presentational skills in staging concerts	References/interview/audition
Ability to motivate other staff and pupils	Choir rehearsal
Ability to organize the work of visiting teachers	Interview/references
Ability to work with external agencies	Interview/references
Knowledge:	
Good subject knowledge at each key stage and post-16	Specialist interview/references
Knowledge of assessment-technique procedures at each key stage and post-16	Specialist interview
Good knowledge of concert repertoire for secondary pupils	Specialist interview
Knowledge of arts in education networks	Specialist interview

Table 8.3 continued

Awareness of how performing arts disciplines can work together successfully	Specialist interview
Knowledge of potential career paths for students and of opportunities for pupils to extend music experience out of school	Specialist interview
Attributes:	
Ability to work with others as part of a team on curriculum issues and in the staging of arts events	Choir rehearsal/references
Ability to lead others, including those from other arts disciplines, on combined arts projects	Specialist interview
Confidence in performing and in conducting or directing the performance of others	Choir rehearsal/audition

Specification. Against each detailed criterion the interviewers would enter a code meaning one of the following:

1. meets criteria in full;
2. meets criteria in the most part;
3. meets criteria in small part only;
4. fails to meet criteria.

This example uses a Music Co-ordinator post in a secondary school because it has sufficient complexity to include most teaching profiles in a school. However, whatever the position being sought, whether in the primary or secondary phase, and whether it is that of headteacher or class teacher, the principle is the same.

Summary

Appointing the right staff means identifying exactly what the needs of the school are. There are six clear steps in moving staffing profiles from shadow to substance. These are:

Candidate Assessment Sheet: Music Co-ordinator

Candidate name: Overall score:

Quality required: *Score (1-4):* *Notes:*

Experience:
Minimum of three years working in a successful music department;
GCSE teaching experience required;
A level teaching experience required;
Experience of performing arts courses desirable.

Qualifications:
Music graduate;
Qualified teacher status.

Skills:
Pianist with good accompanying skills;
Proven organizational skills;
Good presentational skills in staging concerts;
Ability to motivate other staff and pupils;
Ability to organize the work of visiting teachers;
Ability to work with external agencies.

Knowledge:
Good subject knowledge at each key stage and post-16;
Knowledge of assessment-technique procedures at each key stage and post-16;
Good knowledge of concert repertoire for secondary pupils;
Knowledge of arts in education networks;
Awareness of how performing arts disciplines can work together successfully;
Knowledge of potential career paths for students and of opportunities for pupils to extend music experience out of school.

Attributes:
Ability to work with others as part of a team on curriculum issues and in the staging of arts events;
Ability to lead others, including those from other arts disciplines, on combined arts projects;
Confidence in performing and in conducting or directing the performance of others.

Figure 8.4 *Music Co-ordinator: Candidate Assessment Sheet*

- ❑ identifying desired outcomes;
- ❑ identifying the skills and experience profile to reach the outcomes;
- ❑ identifying a suitable staffing structure;
- ❑ writing job descriptions;
- ❑ writing person specifications;
- ❑ creating an interview structure to test for the qualities identified on the person specification.

APPOINTING THE RIGHT GOVERNORS

The responsibilities held by a governing body and the range of tasks they must undertake in order to fulfil these responsibilities are very wide indeed. It is perhaps strange then that in some schools little thought is given to the range of skills needed by the governors in order to fulfil their duties properly.

The skills needed by the governing body

A governing body of a community school is required to have a balance of parent, teaching, staff (non-teaching), LEA governors and co-opted governors. Co-opted governors are appointed by other non-co-opted governors and must also include, if appropriate, representatives from each of the minor authorities (such as a town or district council), the school sponsors (such as a business with a special material relationship with the school) and the Education Action Zone (if the school operates within one). Co-opted governors cannot be people currently employed at the school, pupils of the school, or elected members of the local council. The headteacher would normally also be a governor but may elect not to be so. For voluntary-aided schools there will also be foundation governors.

This structure ensures that the different interest groups concerned with the success of a school are represented. However, it does not relate at all to the actual responsibilities that governors hold and the associated capabilities and knowledge that governors should have. Governors are ultimately responsible for *every* aspect of the running of the school. Yet only in respect of the curriculum are they guaranteed some element of professional input from the headteacher and professional staff of the school. There is no automatic access to financial knowledge, knowledge of personnel and recruitment procedures, knowledge of buildings maintenance and development strategies, contracts for minor works, or even long-term development planning.

While this is not to suggest that each governing body should have an accountant or other financial expert amongst its ranks, there is no denying that – particularly in small schools – life is a lot easier for the headteacher and the rest of the governors if one is available. Similarly, schools should not expect to have a buildings surveyor, architect or other such buildings expert amongst their membership, but where such expertise exists the development of a well-informed maintenance and buildings development plan would be a true gift to the headteacher and other governors.

Knowing where to get help

Not every governing body is going to be able to find specific skills in finance, buildings and personnel from amongst its community. What is important, though, is to find the type of person who is willing to be a point of reference or first contact in these areas of expertise and who is willing to attend training that will help a school in moving issues forward. First-hand expertise is not always necessary as long as people know when to ask for help and where to get it.

Some schools benefit enormously from skilled parents who, though too busy to make a regular commitment over a number of years by serving as a governor, are happy to be involved in a specific project over a number of weeks or months. Training can help enormously in making governors feel comfortable with their heavy and diverse responsibilities. However, training costs money, and hiring professional assistance or advice also costs money. So there has to be a clear commitment on governing bodies that lack internal expertise, or access to volunteers from the wider community, to make provision for this expertise to be bought in from elsewhere.

Getting the balance right

There is often a tendency on governing bodies to fill the ranks up, regardless of the category, with either parents or educationalists or both. There is nothing to stop a governing body doing this. A parent can be appointed as a co-opted governor and a teacher can be appointed as a parent governor.

The problem with this arrangement, however, is that a governing body dominated by parents from the local community and with teachers and lecturers is not likely to bring fresh ideas, innovation or a suitable element of challenge into the school. It is too easy for parents to develop too close a relationship with individual staff, and this relationship can make necessary change later on very difficult. Parents are, understandably, concerned with how their child's school is currently performing – often regardless of the school's position in league tables if their child is happy and appears to be doing well. But those parents may then be reluctant to make changes that they think may affect their child. In most instances this means standing still, or moving forward very cautiously. Sometimes institutions need to take a chance if they are going to move their performance up a gear. Today's parents sometimes find it difficult to think of the needs of tomorrow's pupils.

Governing bodies need to know their schools well, but they also need a good element of detachment – a willingness to accept that the current shape of provision is not set in stone and the ability to see the school in a broader context. This is true vision, unhampered by too close a personal attachment to the institution that they are responsible for developing.

Some of the most innovative and supportive governing bodies have within their membership a balance of parents, educationalists and figures from industry, commerce, law or the health service. Such a balance can prevent a school from becoming too introspective and can introduce an element of external and differential professional influence to the management and working practices in the school. Good organization, ruthless prioritization and a sharp focus on the real issues can serve to keep morale up in a profession that has suffered from low morale in recent years. Governors often underestimate the amount of influence they actually have on a school. In addition to making the working life of the headteacher more bearable through providing support, and bringing expertise and knowledge in some areas to the governing body, their overview of school development issues, from target setting to buildings maintenance, can impact directly and positively on the day-to-day work of teachers.

Building the team

Appointing the right governors is important if the governing body is to have any real influence on the school and is to be of real help to the school. A skills analysis exercise of the type carried out for teachers might be useful, although it can be very difficult for schools in certain areas to find some of the desirable skills from within their own communities. Sometimes large institutions will help, and an approach to banks, building societies and other large organizations can occasionally pay off as some have a policy of encouraging key staff to serve as school governors.

Of more importance to governing bodies is the appointment of governors who will give their energy and time to the task. Time and energy are needed to learn about the school's strengths and needs and to get to know the headteacher or other staff; to attend meetings such as subcommittee meetings, interview panels, exclusions appeals, or parental complaints; to attend training so that the work of the governing body is more easily understood and shared; and to support the headteacher – for example by dealing with outside contractors, getting estimates for work, monitoring the maintenance and condition of the premises, and a host of other vital tasks.

In appointing to a governing body, in order to ensure a balanced and effective contribution from new members, governors should consider the following points:

- ❑ *Match skills against responsibilities.* Take into account the terms of office of existing governors and how long they are likely to remain as governors.
- ❑ *Identify areas needing coverage.* Ensure that key areas concerning finance, recruitment, buildings and maintenance, as well the curriculum and areas such as special needs and exclusions, are all considered.
- ❑ *Consider training versus experience.* Decide whether to opt for a skilled appointment or for more training for an appropriate governor. Consider the likelihood of attracting a professional in certain key areas. The use of parental or local business networks can help in identifying individuals suitable for co-option.
- ❑ *Ensure that the training budget is sufficient to meet requirements.* If expertise cannot be appointed, then those with interests in a particular aspect of school life must be trained up.
- ❑ *Ensure that new governors understand their commitment.* The commitment is to improving the schools. This means not only attending meetings but contributing to them. It means getting to know the school and its work. It means being prepared to attend some training in a particular area of the governors' work.

Knowing the strengths of the existing team will help identify where additional skills or interests are needed. The most important message to give to prospective governors concerns the commitment that they are taking on. If all governors were to understand this commitment at the outset, the workload would be much more bearable for every member of the governing body.

Summary

Appointing the right governors is essential to the success of the governing body in providing valuable support to the school and fulfilling its statutory responsibilities. Remember that governors ultimately hold the final accountability for the success or failure of a school.

A skills analysis of the type used for identifying staff needs, together with a commitment for training, are both helpful strategies. It is also important to keep the balance right amongst the professional parties, the parents and the truly independent parties on the governing body, so that the vision of development is not sacrificed to serving only the needs of the current pupil population.

KNOWING YOUR PRIORITIES

There is so much to consider in the running of a modern school. Prioritization is essential if management, governors and teachers are to keep themselves properly focused on the real issues of school management and school improvement.

Handling the paper mountain

One of the greatest areas of complaint – for headteachers and governors alike – centres on the sheer amount of paper received during the course of a week. Most efficient professionals have ways of dealing with overloaded in-trays. After a quick scan to filter out urgent letters and statutory replies, the pile of pending paperwork can look considerably more manageable. What people tend not to do, however, is to bin the rest! It stays – often for a term, sometimes longer. It grows bigger, being fed every day with the inessential papers arriving, until it becomes enormous.

Its very presence is a negative influence. Almost everybody has had one of these 'pets' sitting threateningly on the corner of their desk! At the end of term the non-essential letters containing 'information' or special offers from school activity centres, or minutes of meetings, together with a mass of attachments, all of which you have no real professional interest in anyway, are often then sorted straight into the bin. This book is not intended to deal in depth with administrative skills, but there are basic rules that can be followed that will help keep a clear desk and will contribute to keeping a clear mind. These are:

- ❑ touch a piece of correspondence once only before deciding what to do with it;
- ❑ deal with it if it can be dealt with quickly;
- ❑ schedule it for later action if it is less urgent or requires a significant amount of dedicated time, but at least decide when you are going to deal with it;
- ❑ delegate it or pass it on if someone else is better qualified or has the necessary skills to deal with it;
- ❑ dump it, particularly if it is inessential, and dumping it will cause no offence or inconvenience to the sender (many circulars and even some meeting minutes will come into this category).

Readers can find more information on managing their desks in such books as *101 Ways to Clean up Your Act* by Dianna Booher (Kogan Page, 1994).

Not losing sight of your aims

The real key to prioritization for a manager or governor is matching the daily workload against the aims of the school. The school development plan has already been identified as the single most important document in the school. That plan should clearly identify the school's priorities and areas for development. The overriding priority for any primary or secondary school is the maintenance and improvement of attainment. The key to achieving this priority is progress. It is this priority that all other priorities support. If pupils are not making progress, attainment is not going to be raised. Priorities within a development plan should ensure that the progress made by the least able child, the average attainer and the high achiever is sustained and improved. If the elements of a development plan do not contribute to raising standards in this way, then the presence of those elements can justifiably be questioned. A good development plan, following the principles outlined in earlier chapters, will serve to keep managers and governors informed of their priorities.

Keeping everyone on target

Governors need to ensure that they not only know their priorities but have recognized the *key* priorities. Teaching staff in addition to managers need to ensure that they know those same key priorities. Time should be set aside at staff meetings and at governors' meetings each term to consider the identified priorities, review the progress the school is making against those priorities and re-prioritize if necessary.

If certain developments become stagnant because they are no longer perceived as being priorities, continued support for them needs to be questioned. It may be a matter of wishing to continue the initiative but diverting resources to a higher priority and delaying the target date for completion to allow for this. Alternatively, it may be a matter of suspending the initiative until resources can once again be properly assigned to it. Or it may be a question of dropping the initiative because it does not contribute to the school's key aims and is not a key priority.

Schools are for learning

In addition to attainment, schools have to ensure the welfare of their pupils. There is much debate as to how much time and resources should be put into this area. A school cannot offer a cure for all the ills of society, and the main priority of a school is to help its pupils develop the broad range of skills, knowledge and understanding that will allow them to succeed in adult life. It is not necessarily a school's job to provide unlimited counselling and care for children with severe problems at home.

A school must decide for itself the balance of pastoral and academic support that it gives to children, taking into account the support available from other sources and remembering that supporting learning is probably what it does best. In a case during the 1990s where two young children were abducted on their way to school, people – including the press and the police – were very quick to point to the fact that their school had 'failed to report the children missing'. To most teachers this must have been disturbing: children are kept way from school every day because of colds, other illnesses, family holidays, and for a host of other reasons. It has always been considered impractical to expect that a school should telephone the parents of each absent child to ask where they are. Society will increasingly move to place a burden of care on those in positions of responsibility. However, an outbreak of flu can result in 50 or more pupils, in a school of over 200, being absent on a single day. No small school could possible check on the whereabouts of each one within the time frame implied by the critics above. Perhaps if parents were required to telephone in their child's intended absence each day before 9.30 am, a call to the police at 10.00 am may be justified when no parental call is received and no pupil has arrived.

The welfare of pupils is nevertheless a factor in ensuring that they come to school ready and willing to learn. Schools need to do enough to ensure that learning takes place but should not be expected to take on the role of other parties or agencies. There is nothing wrong with the idea that appropriate agencies can operate in schools, providing these additional services, particularly after the end of the school day and at breaks throughout the day; but it should not be expected that teaching professionals should provide them or that educational resources should be spent on them.

Priorities, then, must centre on progress. Every activity that the school provides, every lesson that is taught, every assembly where a presentation or talk is given should have, as its primary aim, the development of the skills, knowledge or understanding of every child present. Every plan that the management devises, every strand of the school development plan, every resourcing initiative and every building initiative should have at its root the aim of making learning easier for children in the school. If expenditure of professionals' time or governors' money cannot support these primary aims, the value of the initiatives or activities needs to be questioned.

Summary

It is easy to become distracted from the real issues in today's schools. People must come before paper. The people who really matter in a school are the pupils, and the aims of the school must centre on improving pupil performance. Although schools have a responsibility to provide adequate care for their pupils during school hours, society must not be allowed to distort this duty unreasonably. Schools are for learning, and their main responsibility is furthering the progress made by pupils. Parents and other social agencies have responsibilities that must not be transferred to schools simply because that is where the pupils are at a certain time of the day.

PLANNING AHEAD FINANCIALLY

Planning the school's future spending priorities, budget provision and contingency funding is essential to the stability and sustained development of the school. It is difficult to do this accurately for sustained growth when the reality of public finances means that the detail of a school budget may not be known until the start of a financial year and that that detail may actually involve a reduction of spending from the previous year and not an increase.

The value of your budget may go down as well as up

Each area of the school development plan will indicate how future budget provisions should support the development of the school. However, governors also need to be prepared at times for financial stringency and an overall *reduction* in spending instead of sustained growth. For those schools who have built up reserves in years where there were healthy surpluses, there is less of a problem. However, using reserves to maintain historical provision, without knowing how those reserves will be built up again is not a wise financial strategy. At some point those reserves may run out, and then the consequences will be much more serious as a school has to shift its pattern of spending significantly and in one adjustment from one that it could afford three or four years ago to one that it should have been slowly adapting to over that time. This means that priorities have to be looked at from the bottom up as well as from the top down.

Assuming that the school has the correct curriculum and staffing model to support its aims, a decision needs to be made about what is the minimum resourcing level that can support that model. Minimum staffing levels need to be looked at as well as desirable staffing levels. Minimum training and material resourcing budgets need to be identified as well as desirable ones. This exercise is always better done at a time when it is possible to make a choice rather than at a time when a decision has to be made.

This way of looking at resourcing will give upper and lower parameters to different aspects of a school's resourcing. For example, a primary school may decide that in order meet its aims a minimum of morning-only support from a trained classroom assistant will be provided to reception classes, all Year 1 classes, and Year 2 classes in the term when National Curriculum assessments are carried out. The school may also decide that, if it can be afforded, it would be desirable to provide all-day support from classroom assistants in reception classes and support for two terms for Year 2 classes.

With such an approach, the upper and lower parameters of support are known. With careful planning, the school can protect its reserves and provide additional support to meet priorities when it is as confident as it can be that the minimum levels of provision in future years will not be affected. If the strain on reserves looks as if these levels cannot be indefinitely protected, the school has a little time to review its overall priorities and to see whether resources can be diverted from lower-priority areas into higher-priority areas.

One successful school with £40,000 reserves took the decision to reduce the amount of welfare support in its early-years classes. The governors were able to do this without attracting too much parental criticism because they had identified with the professional staff of the school the minimum level of support that should be provided in these classes. That provision was made known to parents. It was also made known, at a time when reserves were much higher and growing, that additional support was being put in on a temporary basis and may be withdrawn if the school found it could not afford for it to continue.

Compare the foregoing strategy with that of another school with a similar financial starting point.

A school with £50,000 reserves used these reserves year after year to support a staffing complement that neither the curriculum needs nor the income generated from pupil numbers could justify. There was no planned staffing profile, only a notion that historically there had been specialist teaching in one or two subjects, that teachers had always had so much 'non-contact' time, and that support levels had always been at their current level.

The first school is able to protect its reserves and run on a minimum staffing requirement when necessary to meet its aims. The second school will eventually have no reserves and may have to abandon non-contact time for most staff and implement big cuts in both support and specialist teaching provision. Standards in the school may, as a consequence, begin to suffer.

The importance of secure financial planning cannot be underestimated. Spending must support the school's priorities and governors must have a strategy for protecting reserves and gradually reducing expenditure in times of adverse financial pressure. The alternative is to have to accept significant spending reductions, which could damage the infrastructure necessary to support the continued development and growth of the school.

Planning ahead

Planning ahead needs courage and vision. There will be times when it may be tempting to raid long-term savings for a specific project to meet an immediate need or to resource a new idea. Schools need contingency money but provision for a long-term project should not be looked at as a contingency resource or else the long-term aim may never be achieved.

One school desperately wanted to increase its size. Its small town-centre site was inadequate for further development and, such was the rate of overdevelopment of the immediate area, no other sites were available. This did not deter the school from saving towards the cost of this aim and when, unexpectedly, an opportunity came to acquire a stake in a new development, it was able to do so.

For years it must have been very tempting to use the massive resource it had built up in other areas. Sticking to its plans and not losing sight of its vision, even when things looked hopeless, paid off for this school in the end.

The governing body in this example had thought through what the educational benefits of creating a larger school were deemed to be. They did, it is true, want a larger school because there were many within their community whom they would like to serve but whom they could not. However, they also recognized that a larger school meant more resources for books, computers, specialist teachers and other enhancements. When 92 per cent of a school's budget is spent on fixed costs (teaching and accommodation services), whatever the size of the school, governors are better off having 8 per cent of £700,000 spare for resourcing and enhancements rather than 8 per cent of £350,000.

Key factors when planning ahead are to:

- ❑ define your vision/long term plans;
- ❑ identify the steps you need to take to achieve them;
- ❑ know the financial cost of your vision over the long term;
- ❑ plan how you are going to support the development of the finance base;
- ❑ don't be deterred or distracted even when problems look insurmountable.

Summary

Education budgets are not, and probably never will be, stable. 'Formula' funding means that there will always be some winners and some losers because costs within the budget vary year by year. It is important to plan the parameters of expenditure needed to support the school's aims and never lose sight of the long-term vision seen for the school.

SETTING TARGETS AND IMPROVING PERFORMANCE

The path to improvement is a difficult one unless there are way-points that can be recognized and attained. Good schools are not simply there to maintain the existing performance of pupils and staff – few schools, if any, are in the position where there is no room for improvement. Even the best selective schools with the smallest classes and the most generous resources need to ask themselves whether or not they are doing enough for each and every one of their pupils.

Where do we go when we are already at the top?

The 'coasting school' is a common phenomenon. A high-ability intake requires little maintenance to produce an above-average output. But the output has to be measured, not only by attainment but by progress, as has been discussed in earlier chapters. Imagine the potential attainments of the most able children if rigorous and concerted efforts were put into extending them beyond what a school had previously achieved with pupils of this category. For many schools, the concept of performance beyond the best of what they had previously achieved is not seen as a viable aim. There is a belief that because they are above the average, and because most parents think that they do well by their children, and because teachers plan and teach for different levels of ability, the school is doing all it can.

Many would argue that any improvement in attainment in good schools is dependent on factors such as increased levels of resourcing, smaller classes, or perhaps more teachers. However, increased learning is not only dependent on a sufficiency of resourcing and teachers. Big improvements in learning can come from exploring different and more efficient ways of learning. One of the big disappointments of the introduction of the National Curriculum was the complete lack of any reference to learning styles, techniques or strategies. Sadly, the National Curriculum concentrated solely on *what* to learn rather than *how* to learn. Subject content isn't that important in the longer term. Of course, children have to learn to read and write and they have to learn basic numeracy skills, but beyond these core skills the subject content is often irrelevant and the stress in trying to teach a young child about materials and their properties or coastal erosion is often an obstruction to the very learning that should be taking place.

It's not what you learn, it's the way that you learn it

Some schools are beginning to teach children memory techniques, planning strategies and stress-management strategies. These schools recognize that learning depends on more than a well-planned and well-presented lesson. Such a lesson may currently qualify as good teaching, but even when 'appropriate tasks' are set for the pupils this does not necessarily translate into good learning.

Learning depends on the pupils' abilities to concentrate, remember, analyse and discriminate. Whereas all of these processes are essential in every subject of the curriculum and many activities are designed within the teaching of the curriculum to help develop these processes, sufficient attention is not often given overtly to teaching children the core, common skills that can be used to support learning across the whole curriculum.

A young music teacher was given a job in an inner-city school where the pupils had 'seen off' the previous three incumbents. There were no resources to speak of and the music curriculum was going to be singing- and listening-based out of necessity. However, the pupils would not sing, much preferring to make a joke out of the activity, and would not settle down for long enough to listen to a piece of music. If the piece of music played was outside their previous experience, they would rubbish it within seconds of its starting.

Rather than give up, the teacher decided to use his own knowledge of relaxation and mind-calming techniques to get them receptive to learning in music. As a pianist, he had been taught to tense and relax, in sequence, each muscle in the hand, wrist, forearm, shoulder and neck in order to produce a relaxed and mentally controlled state before performing. As a drama and yoga student he had been taught a similar technique involving the whole body in order to produce a state of extreme calm and peace but also increased awareness.

The teacher led the class in performing these or similar techniques at the start of each lesson. The activities were so different to anything that the pupils had experienced before that they went along with them. The results were spectacular. Ten minutes relaxation and focusing the mind and they would concentrate on learning new songs, on improving their singing techniques, and even on listening to new music. The pupils said they enjoyed the lessons because they felt good. The actual musical content was no different from what had originally been intended. However, these pupils were being taught to relax and to focus their concentration with immediate benefits to learning in music.

The teacher survived and established a thriving music department at a school that had previously been deemed as a no-go area for music teachers.

If it is possible to get these kinds of results with highly reluctant learners, imagine what could be achieved through teaching techniques such as these to other pupils. The point here is that in our educational thinking we have hardly begun to scratch the surface in what it is possible for our children to learn – and we never will do so as long as we are tied to yesterday's techniques, yesterday's curriculum and yesterday's lack of vision. Yesterday's methodologies have served us in attaining an acceptable level of intellectual and technical developmental for the mid to late part of the 20th century. However, hanging on to them, and not recognizing that to reach new heights of intellectual and technical attainment we need new ways of learning, will consign our country to sliding further and further down the international league table of educational performance. If schools, especially good schools, are to raise standards further, they must look beyond resourcing and staffing levels and equip themselves with new strategies to improve learning.

Stretching the horizons

The setting of targets is essential in the school improvement process. Target setting is now mandatory, of course, and schools have to identify targets in literacy and numeracy at the very least. High-achieving schools can ask the question, 'Where do we go when we already at the top?' The answer is staring them in the face, but the very structure imposed through National Curriculum levels is often the biggest obstacle to understanding the answer.

Schools that ask this type of question often attain 95 per cent or more at Level 2 in English and mathematics at the End of Key Stage 1 and may attain 20 per cent or more at Level 3. They may have similarly good attainment at the End of Key Stage 2. This fits in with the nationally recognized criteria for a successful school in performance terms, and the structure of the National Curriculum implies that schools such as these are doing more than enough already. The challenge for these schools is to do something radical: to shift the thinking of the school up a gear, to ignore national norms, and to aim to establish a new average with more and more children attaining Level 3 and more and more moving towards – or even attaining – equivalent to Level 4.

A radical approach will require looking at the way pupils learn and a shift in expectations. Teachers will have to be taught new learning skills and will have to be convinced that they can work with young children. Schools will have to 'dare to be different' and teach disciplines that enhance memory, improve concentration, and encourage analysis and discrimination.

In the same way that these strategies apply to 'successful' schools, they also apply to schools that genuinely have an average, or even a below-average, intake. Nobody need be afraid of targets. It is easy to assess the current and historical levels of attainment and say that this is the best we can hope for in the future – which may well be the case, given that nothing else changes in the school. However, target setting isn't only about pushing the children to their limits using the existing repertoire of curriculum structure, content and teaching techniques in the school. Increase the targets without changing the strategies and there will come a time when existing techniques alone have no more impact on performance. The result is often stress for both teachers and pupils, who find it increasingly difficult to achieve higher and higher targets. Much as in the same way that new, and sometimes radical, technology in industry has resulted in improved efficiency and performance, new strategies have to be used in the classroom to improve learning and raise standards.

Target setting isn't, then, only about aiming for a number slightly higher than the last number and going for it. That technique will work to a point, but it will only succeed in taking up the slack in existing teaching and learning strategies. Beyond that, the real prizes are for the schools that teach learning before they teach history and geography, concentration before literature and music, analysis and discrimination before science and technology.

Summary

An attempt to define average standards for pupils' attainment means the introduction of an artificial ceiling of expectation for some pupils. Current methodologies are very much taken for granted. Big improvements in attainment will come, in future, not from defining what should be learned but through national initiatives on how learning should take place. Attention should be given to opening up the full potential of the human brain rather than simply serving and reinforcing the potential we already know.

KNOWING YOUR SCHOOL'S STRENGTHS AND WEAKNESSES

Schools can only improve if they have accurate information on their current performance. They can only have this if they are capable of reviewing and evaluating their own success and identifying their own weaknesses. Pulling together the wealth of performance-related information that a school has in its possession means that the information can be used effectively in raising standards.

Consider your knowledge base

So what information does the school have at its disposal? Much of it we have discussed in detail in earlier chapters, but now let us consider it in its totality and the part that each measured area plays in influencing the outcome of the whole if the messages are interpreted properly and acted on quickly.

The following is a list of seven data types:

- *Pupil performance data*. This can be broken into two categories. Firstly, there is the statutory, national category. In a primary school these are End of Key Stage assessment tests, while in secondary schools they include GCSE and other accredited examination systems. Secondly, there is the non-statutory but well-structured in-house category. These include standardized tests for reading and writing and any internal regular, monitored and moderated system for assessment.
- *Progress indicators*. These are derived from tracking pupils' performance at individual, class and year levels over a period of time. Their accuracy relies on consistency in test conditions and type over time. Progress indicators can also be used to measure performance year on year at a particular stage in the education process. This is the strategy used by the government to tell us whether or not standards generally are rising or falling. Schools dislike them because they do not take into account the attributes of the particular cohort and therefore they are of limited value in assessing the progress made by those children.
- *Teacher performance data*. Headteachers are now being given information on how Ofsted inspectors judge the teaching capabilities of each individual teacher in a school. This information should not be used on its own to judge a teacher's performance. Managers should be developing written records of regular observations, using established criteria, carried out on every teacher, as well as the targets that follow those observations and the progress made against those targets.
- *Attendance and punctuality data*. Many schools now analyse this data to see if they can identify target groups who may have common factors influencing their attendance patterns. In good schools this can lead to a modified curriculum or extra support, in addition to rigorous monitoring, attendance targets and increased levels of contact with home.
- *Behaviour data*. Schools that have a good and consistently used system for rewarding good behaviour and challenging poor behaviour have the capacity to identify trends and groups of pupils where additional support or attention may be required. The system

only works if all staff use it and if written reports conform for easy analysis and are collected centrally. Schools can identify *improving* behaviour and attitudes as well as problems.

- ❑ *Financial and resourcing data*. Many schools have good financial monitoring systems but do not make the connection between spending and measuring the effect of that spending on pupil performance.
- ❑ *Management systems*. Communication and decision-making processes should be clearly defined, and it should be possible to track back the decision-making and consultation process for any initiative. Minutes of meetings should be kept and should be accessible. Monitoring and evaluation strategies should be known to all staff, and findings in these areas should be shared appropriately as soon as possible.

The systems described above interrelate as shown in Figure 8.5.

A well-organized school that is using the systems listed above and shown in Figure 8.5 to promote learning, progress and attainment will be a successful school. If staff are kept informed and involved, and are challenged as well as given recognition for their work, then morale will remain high. Conversely, it is the absence of these systems, or their misuse, that weakens the credibility of a school and lowers the morale and performance of all who work in it.

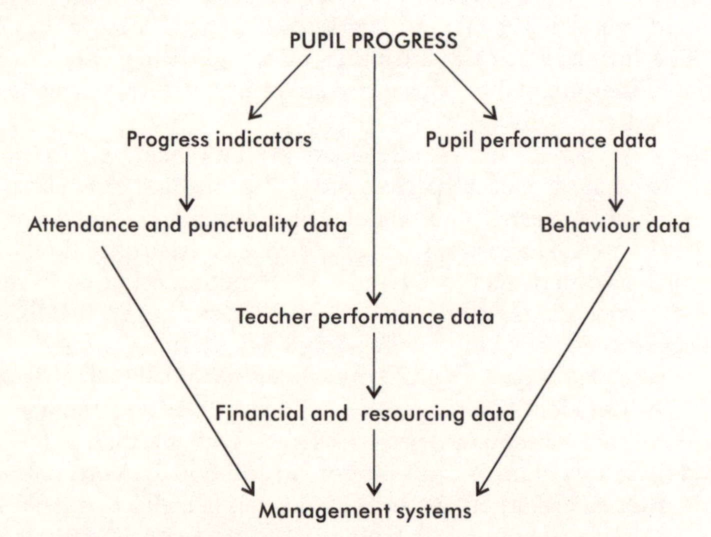

Figure 8.5 *The interrelation of data and systems*

WORKING TOGETHER FOR SUCCESS

A school may have devised potentially good systems for everything it does and may have, on paper at least, an impressive array of policies and procedures for just about everything that it is involved in. However, plans on their own mean nothing unless every part of the organization is working with the rest to make these plans succeed. A well-written school development plan is worthless if it was only written by one person and no one else has been involved. School policies are meaningless if the staff and governors do not have shared ownership of them. Even efficient financial management is worth relatively little if all it does is keep the books tidy for their own sake and if financial decisions are not made with educational improvements in mind.

Practice to paper

In the early days of national school inspections, schools were often obsessed with having a policy for everything. This despite Registered Inspectors saying to them before the inspection that if paperwork was missing or underdeveloped it was better to leave it that way rather than rush something together. There was not a full appreciation that the content of school documentation is often one of the very criteria that inspectors use to judge the efficiency and effectiveness of management.

On a pre-inspection visit to examine school documentation, an inspector found a sentence at the start of a policy that read 'Here at xxxx Primary School we pride ourselves on giving all children access to information technology.' The problem was that the school being inspected was actually yyyy Primary School! An extremely embarrassed headteacher, when faced with this apparent contradiction, said that he felt very let down by the staff member responsible. . .. Hang on a minute, though: aren't policies supposed to be in the ownership of the whole school, including the management and governing body? Surely a policy shouldn't be a policy if only one teacher had read it. Why did the headteacher allow it to be tabled as his school policy?

Needless to say, the inspection went downhill from this point on. At the management and governors' feedback sessions, it was clear that this was not a united school. Excuses and finger pointing lent weight to the inspectors' findings that the school was poorly managed and poorly governed. Hundreds of children had been ill-served by this institution and dozens of teachers had risked having their careers potentially wrecked by working for such slack management.

Pulling together

It is clear that, in order to succeed, everybody in the school has to be working towards the same aim. Governors need to work well together so that their workload is shared out properly and so that they all know what the school's current performance is, what their priorities are, and what needs to be done to achieve them. Governors need to work with management and with teachers on subcommittees and working groups so that the professionals feel that they have the support and guidance of those who are ultimately responsible for the school and its performance. The management team needs to work together to give clear and consistent leadership to staff. Staff are responsible for working within the structures and frameworks adopted by the school and, through their work, contributing to the aims of the school. This interrelation of the various groupings is shown in Figure 8.6.

It is sometimes apparent that relationships between some members of staff are weak, or that there are imbalances of power within a school. For example, a common failing in weak primary schools is the absence of any kind of meaningful professional partnership between the headteacher

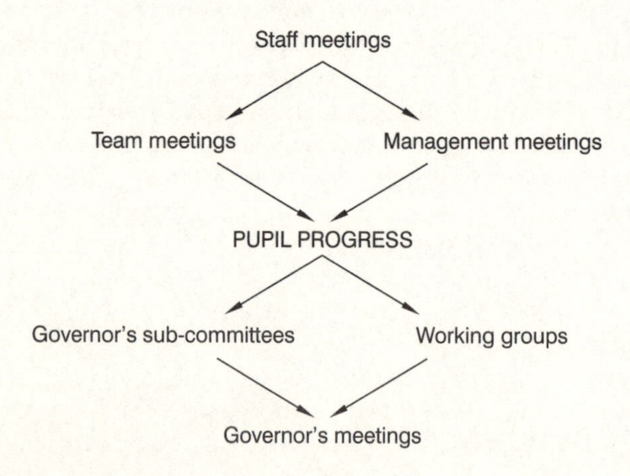

Figure 8.6 *The interrelation of a school's groupings*

and the deputy head. Likewise, it can be disturbing for an individual teacher to know that some staff, for no professional reason are more involved, favoured and have more influence than others. It is disturbing if one teacher is supported in an initiative and another is not, when no explanation or rationale is given for the inconsistency. Schools are not cosy social institutions where the patronage of the headteacher counts for everything. They are professional places of business where the priority of the day is educating children. There is no time for personal prejudices or favouritism to get in the way of educational priorities. Staff need to work together to ensure that they follow the school policies, especially in respect of pupil management, as it is very unsettling for pupils to be faced with inconsistent responses from different teachers.

Keeping the focus clear in meetings

When setting up meeting structures, it is vital to ensure that there is a good flow of relevant information from one level of management to the next and that decisions or initiatives emanating from these meetings are properly disseminated to all staff. Whoever is responsible for putting together the agenda for particular meetings should ensure that the agenda reflects the priorities of the school. Work that can be done outside of a meeting structure by individuals or small groups should not take up other teachers' time by doing it within the meeting. It may be enough for tasks to be delegated and the results brought back to a future meeting for discussion.

Each meeting should have clear aims, with a short list of identified target outcomes appearing somewhere within the agenda. The foot of the agenda is a useful position for this rather than the head because the course of the meeting will not be sidetracked by decisions to come if the target outcomes are at the end; furthermore, the outcomes themselves will act as an indicator as to the success of the meeting. This isn't to say that outcomes should be predetermined, but is simply a reminder that a decision on an issue – one way or the other – has to be reached, and that an aim of the meeting is to reach that decision. If the outcomes that a meeting produced do not have the potential to improve standards, the time spent in the meeting has probably been wasted.

It is also useful to have as many stakeholders as possible involved in the decision-making processes of the school. Many schools develop initiatives within small management groups, take them to staff meetings, refine them and take them to governors' meetings, sometimes referring them to a governing-body subcommittee on the way. Good practice involves getting classroom teachers and governors involved at an early stage so that initiatives are not solely management-driven and so that ownership is more widely spread. Wider ownership of an initiative almost always guarantees greater levels of success.

It easy for managers and governors, and for teaching staff, to get locked into cycles of meetings that become more and more remote from the real business of schools. If every proposal, every agenda item and every initiative were to be prefaced with a statement saying how it would contribute to the educational development of pupils in the school, then relevance would be guaranteed.

Summary

The importance of all groupings within a school working together as a team cannot be underestimated. This applies to teachers, managers, other staff and governors in ensuring that the policies produced for the school reflect both best practice and high aspirations.

The aim of any meeting or other action in a school is to further the attainment of pupils through helping them make progress. Sometimes it is easy to forget this or take it for granted. People must keep a clear focus on the true aims of the school, both in their individual actions and when interacting with each other in meetings. In this way the interests of the children will always be served.

BUILDING A SUCCESSFUL SCHOOL

All schools can be successful. High standards will come automatically for many schools if progress is good, and for some if residual pupil ability is high. But high standards alone do not make a school successful. A recurrent theme throughout this book has been that true success is defined in the progress that pupils make in their learning.

Good management will identify the factors influencing pupil progress and will concentrate on addressing those factors. This means knowing the school intimately and promoting a complementary ethos and mode of operation that will build on its strengths and eliminate its weaknesses. Knowing the backgrounds of the pupils will help influence teaching and learning methodologies and will help shape reward-and-sanction strategies. Knowing the strengths and weaknesses of the teachers will help focus training and support where it is necessary and where it will have the most impact in promoting pupil progress. Knowing the abilities of pupils will help teachers plan their work more effectively and promote more rapid learning.

Effective schools need to remain effective. This means having the systems in place that will draw attention to any emergent weaknesses in learning, teaching, management and governance. Successful schools understand accountability from the top down and always take responsibility for identified weaknesses. Good monitoring systems provide valuable information. This information is essential in providing governors, managers and parents with the reassurance that the school and its pupils are performing well, or to give early warning that something is amiss. Good systems will provide consistency and security for teachers and pupils alike. Effective and clear systems give those working within them confidence to operate within known parameters. Confident workers will often be more successful workers.

Successful schools will be selective about where they put their energies. Learning is not simply about acquiring content-based knowledge. Effective learning is about developing techniques that promote independence in learning; improved memory and understanding and the ability to connect and relate learning in one context to problems encountered in another. Schools that want to be on top will place a high priority on learning because they know that curriculum knowledge will then be acquired much more easily.

Above all, successful schools will have vision. They will understand that a changing society will require changes in their priorities and in the way they operate. They will have ambition for the school that will be clearly defined. Staff and governors will work together in formulating common aims that support the development of pupil learning – the core function of the school, after all – and in realizing them.

A school is ultimately successful when managers, governors and teachers work together effectively to maximize the learning of every child in their school. To do this they need an understanding of the key elements in building a successful school. These are:

- ❑ understanding true success;
- ❑ understanding the factors that influence success;
- ❑ knowing the context that the school is working in;
- ❑ knowing the strengths and weaknesses of the staff;
- ❑ knowing the abilities and levels of knowledge of the pupils;
- ❑ having effective systems to inform management and guide working practices;
- ❑ knowing the real priorities and promoting effective learning;
- ❑ having vision – keeping an eye on the ball and knowing where the goal is.

CHAPTER SUMMARY

This final chapter has tried to emphasize the importance of integration in all the activities of running a school – integration of personnel selection, of systems, of working groups and of aims. The main points arising from the chapter are as follows:

- ❑ It is important to have clear strategies for strengthening weak schools and for ensuring a secure and successful future for all schools.
- ❑ A school must know its staffing priorities and appoint the right staff for vacancies that occur – at whatever level.
- ❑ It is better to extend the appointment of good supply teachers until the best permanent staff are secured rather than settle for a quick appointment.
- ❑ Governors should identify their strengths and actively recruit people to their governing body who can offer different skills and who are prepared to pull their weight.
- ❑ Strategic development planning is essential, as is having measurable aims.
- ❑ Target setting, particularly in the context of the high-achieving school, means challenging the norm and looking beyond the existing repertoire of curriculum content and traditional teaching and learning techniques.
- ❑ Schools can only know how to develop if they are capable of reviewing and evaluating their own success and identifying their own weaknesses. All functions of a school must support learning by pupils.
- ❑ The basic principles of good management include having clear aims, an ability to prioritize, good communication, high expectations of teachers, consistency, positive and supportive leadership, and very good organizational skills.
- ❑ It is essential that the headteacher, teachers and governors work together in a school with the educational needs of each individual pupil at the forefront of their thinking and actions.
- ❑ A clear understanding of the key elements in building a successful school is vital.

INDEX